Mathematics for
the Physical Sciences

Springer
*New York
Berlin
Heidelberg
Barcelona
Hong Kong
London
Milan
Paris
Singapore
Tokyo*

James B. Seaborn

Mathematics for the Physical Sciences

With 96 Illustrations

 Springer

James B. Seaborn
Department of Physics
University of Richmond
VA 23173, USA
jseaborn@richmond.edu

Library of Congress Cataloging-in-Publication Data
Seaborn, James B.
  Mathematics for the physical sciences / James B. Seaborn.
      p.   cm.
  Includes bibliographical references and index.
  ISBN 0-387-95342-6 (alk. paper)
    1. Mathematical physics.   I. Title.
  QC20.M364   2001
  530.15—dc21   2001049271

Printed on acid-free paper.

Production managed by Jenny Wolkowicki; manufacturing supervised by Jerome Basma.
Typeset by The Bartlett Press, Inc., Marietta, GA.
Printed and bound by Maple-Vail Book Manufacturing Group, York, PA.
Printed in the United States of America.

9 8 7 6 5 4 3 2 1

ISBN 0-387-95342-6   SPIN 10847909

Springer-Verlag   New York Berlin Heidelberg
A member of BertelsmannSpringer Science+Business Media GmbH

To my friend, Matthias Wagner

Preface

This book is intended to provide a mathematical bridge from a general physics course to intermediate-level courses in classical mechanics, electricity and magnetism, and quantum mechanics. The book begins with a short review of a few topics that should be familiar to the student from a general physics course. These examples will be used throughout the rest of the book to provide physical contexts for introducing the mathematical applications. The next two chapters are devoted to making the student familiar with vector operations in algebra and calculus. Students will have already become acquainted with vectors in the general physics course. The notion of magnetic flux provides a physical connection with the integral theorems of vector calculus. A very short chapter on complex numbers is sufficient to supply the needed background for the minor role played by complex numbers in the remainder of the text. Mathematical applications in intermediate and advanced undergraduate courses in physics are often in the form of ordinary or partial differential equations. Ordinary differential equations are introduced in Chapter 5. The ubiquitous simple harmonic oscillator is used to illustrate the series method of solving an ordinary, linear, second-order differential equation.

The one-dimensional, time-dependent Schrödinger equation provides an illustration for solving a partial differential equation by the method of separation of variables in Chapter 6. Two examples are discussed in detail—one from quantum mechanics (the quantum harmonic oscillator) and the other from classical electrostatics (a conducting sphere in a uniform electric field). In both cases, physical boundary conditions are used to constrain the parameters introduced in the equations. A more general discussion of boundary value problems is given in Chapter 7. Two more examples are considered in some detail—again, one from classical physics (a vibrating drumhead) and one from quantum physics (a particle in a one-dimensional box).

Orthogonal functions are treated in Chapter 8. The chapter opens with a brief discussion of the failure of classical physics beginning around 1900. The need for a new physics provides the motive for discussing mathematical operators, eigenvalue equations, and the virtues of orthogonal functions. The quantum harmonic oscillator illustrates the formulation of an eigenvalue problem and the appearance of the discrete nature of physical quantities in quantum physics. This treatment of the quantum harmonic oscillator is then seen as a special case of orthogonal functions arising in the larger context of Sturm-Liouville theory. The usefulness of orthogonal functions is further illustrated in the remainder of this chapter in connection with Fourier integrals, Fourier series, and periodic functions applied to problems in both classical and quantum physics.

The matrix formulation of an eigenvalue problem is introduced in Chapter 9 with a view toward laying a mathematical foundation for Heisenberg's matrix mechanics. Two applications from classical physics are also considered—a system of coupled harmonic oscillators and the principal axes of a rotating rigid body. A brief discussion of variational calculus and a derivation of the Euler-Lagrange equation are given in Chapter 10.

A set of exercises is provided at the end of each chapter to give the student experience in applying mathematics to problems in physics. Illustrative examples are worked out in the text. A number of exercises call for graphical representations of the solutions. Some are particularly amenable to solution by numerical methods. Most science and mathematics students have some expertise in using software systems such as *Mathematica* and *Maple* or plotting routines like *Physica* and will be able to apply their computational skills to these exercises. However, to avoid giving the impression that computers are necessary for the book to be useful, I have not treated numerical and computational methods in the text.

The purpose of the book is to collect in one place some essential mathematical background material commonly encountered by undergraduate students in upper-level physics and chemistry courses. There is nothing new here, of course. There are a number of well-written, comprehensive texts on mathematical methods in physics that offer much more extensive treatments of this material along with a wide range of other mathematical topics. All of this material is also treated to some extent in various upper-level undergraduate physics texts, but to my knowledge not all of it in a single book. I have listed in the References those works that have been most helpful to me in putting together this one. For the reader who wants more detail or a different view, some additional sources are recommended in a short Bibliography.

University of Richmond   James B. Seaborn
July 2001

Contents

1
A Review

Our aim in this book is to introduce you to some mathematical applications that you will encounter in intermediate and advanced undergraduate courses in classical mechanics, electromagnetism, and quantum physics. To provide physical contexts in which to set these applications, we shall draw on a few familiar concepts from your general physics course. In this first chapter, we review these concepts, work some examples, and give you some exercises at the end of the chapter to help refresh your memory about them.

1.1 Electrostatics

An electric field is said to exist in a region in which an electric charge experiences a force. The force $\mathbf{F}$ exerted on an electric charge q' in an electric field $\mathbf{E}$ is[1]

$$\mathbf{F} = q'\mathbf{E}. \tag{1.1}$$

According to Coulomb's law, the force $\mathbf{F}$ that an electric charge q exerts on a second charge q' is[2]

$$\mathbf{F} = \frac{kqq'}{r^3}\mathbf{r}, \tag{1.2}$$

---

[1] Boldface type is used to denote vector quantities (e.g., $\mathbf{F}$ for force) and italic type for scalar quantities (e.g., q for electric charge). A more detailed discussion of vectors will be given in Chapters 2 and 3.

[2] In mks units, the Coulomb constant has the value $k = 9 \times 10^9$ N·m²/C².

where $\mathbf{r}$ is the vector that gives the position of q' relative to q.

From Eqs. (1.1) and (1.2), we see that the electric field due to a single point charge q is

$$\mathbf{E} = \frac{kq}{r^3}\mathbf{r}, \tag{1.3}$$

where $\mathbf{r}$ gives the position of the field point relative to the source q. The superposition principle holds so that the electric field due to a distribution of point charges is equal to the vector sum of the fields due to the individual charges. For N charges, we have

$$\mathbf{E} = \mathbf{E}_1 + \mathbf{E}_2 + \cdots + \mathbf{E}_N = \sum_{i=1}^{N} \mathbf{E}_i.$$

Example 1.
Two electric charges are placed 2.30 cm apart at the vertices of a right triangle, as illustrated in Fig. 1.1. Calculate the resultant electric field at the point P at the (vacant) right-angle vertex of this triangle.

Solution.
The direction of an electric field is the direction in which a small positive charge experiences a force. The directions of the fields $\mathbf{E}_1$ and $\mathbf{E}_2$ due to charges labeled q_1 and q_2, respectively, are shown in Fig. 1.1. The rectangular components of the resultant field $\mathbf{E}$ are seen from Fig. 1.1 to be

$$E_x = -E_2 = -\frac{(9 \times 10^9 \text{ N} \cdot \text{m}^2/\text{C}^2)(4.96 \times 10^{-9} \text{ C})}{(0.023 \text{ m} \cos 28°)^2} = -1.082 \times 10^5 \text{ N/C}$$

$$E_y = E_1 = \frac{(9 \times 10^9 \text{ N} \cdot \text{m}^2/\text{C}^2)(2.37 \times 10^{-9} \text{ C})}{(0.023 \text{ m} \sin 28°)^2} = 1.829 \times 10^5 \text{ N/C}.$$

From these components, we find that the magnitude of the electric field at P is

$$E = \sqrt{E_x^2 + E_y^2} = 2.12 \times 10^5 \text{ N/C}$$

and the angle that $\mathbf{E}$ makes with the positive x-axis is

$$\tan^{-1} \frac{E_y}{E_x} = 120.6°.$$

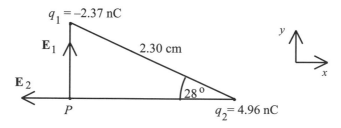

FIGURE 1.1. Charge distribution for Example 1.

Thus, the resultant electric field at the point P is 2.12×10^5 N/C up to the left at $59.4°$ above the horizontal.

1.2 Electric Current

In certain materials, especially metals, electric charges are free to move. Applying an electric field to the material causes the electric charge to flow, hence an electric current is produced. Electric current I is defined as the time rate of flow of electric charge, that is

$$I = \frac{dq}{dt}. \tag{1.4}$$

Example 2.
A copper wire carries an electric current of 2 amperes. How much electric charge passes through the wire each minute?

Solution.
This problem is simply a matter of unit conversion,

$$I = 2\,\text{A} = 2\left(\frac{\text{C}}{\text{s}}\right)\left(\frac{60\,\text{s}}{1\,\text{minute}}\right) = 120\,\text{C/min}.$$

A charge of 120 coulombs flows through the wire each minute.

Example 3.
A straight beam of electrons is incident on a fixed target. The beam has a uniform cross section of diameter 0.10 mm. The average particle density in the beam is 4.5×10^{14} electrons/m^3 and the corresponding electric current is 3.2 μA. Calculate the average speed of the electrons in this beam.

Solution.
The beam current I can be represented in terms of the amount of electric charge ΔQ passing a given point (indicated by the vertical arrow in Fig. 1.2) in a time Δt. From Fig. 1.2, we see that ΔQ is equal to the electric charge density multiplied by the volume of the shaded beam element of length $v\Delta t$. Thus,

$$I = \frac{\Delta Q}{\Delta t} = \frac{en\pi\left(\frac{d}{2}\right)^2 v\Delta t}{\Delta t} = en\pi\left(\frac{d}{2}\right)^2 v, \tag{1.5}$$

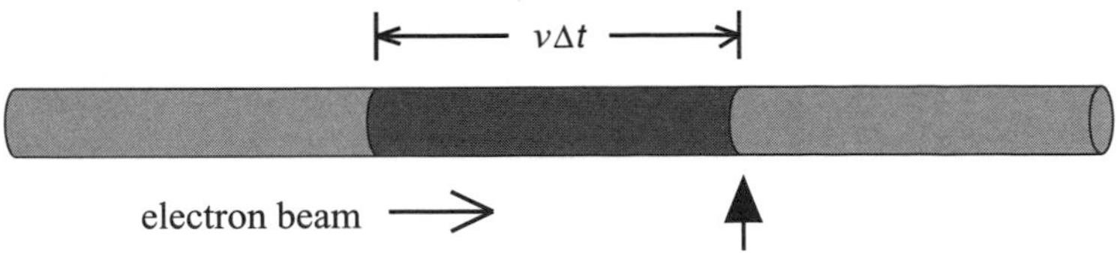

FIGURE 1.2. Electron beam for Example 3.

where e is the electric charge of the electron, n is the particle density of the beam, d is the diameter of the beam, and v is the average particle velocity. Rearranging Eq. (1.5), we find that the particle velocity is given by,

$$v = \frac{I}{en\pi \left(\frac{d}{2}\right)^2} = \frac{3.2 \times 10^{-6}\,\text{A}}{(1.6 \times 10^{-19}\,\text{C/elec})(4.5 \times 10^{14}\,\text{elec/m}^3)\pi \left(\frac{10^{-4}\,\text{m}}{2}\right)^2}$$

$$= 5.6 \times 10^6\,\text{m/s}.$$

Hence, the average speed of the electrons in the beam is 5.6×10^6 m/s.

1.3 Magnetic Flux

According to Faraday's law, the emf (voltage) V induced in a conducting loop is equal to the time rate of change of the magnetic flux Φ enclosed by the loop. That is,

$$V = \frac{d\Phi}{dt}.$$

For a coil consisting of N identical loops, the total emf induced in the coil is

$$V = N\frac{d\Phi}{dt}. \tag{1.6}$$

Magnetic flux (a scalar) is defined in terms of the magnetic field vector **B** (often called the *magnetic flux density*) according to

$$d\Phi = B_\perp dA,$$

where $B_\perp$ is the component of **B** that is perpendicular to the infinitesimal area element dA. The total magnetic flux Φ enclosed by the conducting loop C in Fig. 1.3 is

$$\Phi = \int_S B_\perp dA, \tag{1.7}$$

where S is any surface bounded by C.

Example 4.

A uniform magnetic field is perpendicular to the plane of a circular coil consisting of 350 identical turns of copper wire. The diameter of the coil is 19.3 cm. The magnetic field is switched off and falls to zero in 2.43 milliseconds thereby inducing an average emf of 2.78 volts. Calculate the original strength of the magnetic field.

Solution.

For the surface S, we choose the plane of the coil. The average emf $\bar{V}$ induced by a finite change in the flux is then given by

$$\bar{V} = N\frac{\Delta\Phi}{\Delta t} = N\frac{(\Delta B)A}{\Delta t} = N\frac{B\pi\left(\frac{d}{2}\right)^2}{\Delta t},$$

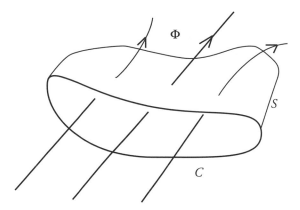

FIGURE 1.3. Magnetic flux through a conducting loop.

where d is the diameter of the coil. Substituting the given values for the parameters, we have

$$2.78 \text{ v} = 350 \frac{B\pi \left(\frac{0.193 \text{ m}}{2}\right)^2}{0.00243 \text{ s}},$$

from which we get $B = 6.60 \times 10^{-4}$ T.

1.4   Simple Harmonic Motion

A particle that moves under the influence of an elastic restoring force is said to obey Hooke's law,

$$F = -kx, \tag{1.8}$$

where F is the force tending to restore the particle to its equilibrium position (at $x = 0$), x is the displacement from equilibrium, and the parameter k gives a measure of the strength of the force required to displace the particle a given distance from equilibrium. From the force, we can calculate the work done in displacing the particle, hence its elastic potential energy,

$$PE = \tfrac{1}{2}kx^2. \tag{1.9}$$

If the particle is moved a distance b away from equilibrium and released from rest, it will oscillate back and forth with a constant amplitude[3] b, provided frictional effects are negligible. In this case, the particle is said to execute simple harmonic motion with a total constant energy that is the sum of the kinetic energy and the elastic potential energy,

$$E = KE + PE = \tfrac{1}{2}mv^2 + \tfrac{1}{2}kx^2, \tag{1.10}$$

---

[3]The *amplitude* of the oscillation is the maximum distance from equilibrium reached by the particle during its motion.

where m is the mass of the particle and v is the particle's speed. The period of oscillation T, i.e., the time required for one complete oscillation of the system, is related to the mass of the particle and the force constant k by

$$T = 2\pi\sqrt{\frac{m}{k}}. \qquad (1.11)$$

Probably the most familiar example of a simple harmonic oscillator is that of a small object attached to an elastic spring.

For a particle released from rest at time $t = 0$ when it is at $x = b$, its position x at an arbitrary time t is

$$x = b\cos\frac{2\pi t}{T}. \qquad (1.12)$$

The corresponding velocity v and acceleration a are

$$v = -\frac{2\pi b}{T}\sin\frac{2\pi t}{T}, \qquad (1.13)$$

$$a = -\frac{4\pi^2 b}{T^2}\cos\frac{2\pi t}{T}. \qquad (1.14)$$

Example 5.
A 12-g mass attached to an elastic spring is pulled 22 mm away from its equilibrium position and released from rest. The mass is constrained to move horizontally with simple harmonic motion making a complete oscillation every 85 milliseconds. Calculate the maximum speed of this mass during its motion.

Solution.
First, calculate the force constant k from Eq. (1.11),

$$0.085\ \text{s} = 2\pi\sqrt{\frac{0.012\ \text{kg}}{k}},$$

which yields a value $k = 65.6$ N/m. Since the mass is released from rest, the total energy is obtained from Eq. (1.10) with $x = 22$ mm and $v = 0$. That is,

$$E = \tfrac{1}{2}(65.6\ \text{N/m})(0.022\ \text{m})^2 = 0.0159\ \text{J}.$$

We also see from Eq. (1.10) that the particle reaches maximum speed v when $x = 0$ (i.e., at equilibrium). In this case, all of the energy is kinetic energy, and Eq. (1.10) gives

$$0.0159\ \text{J} = \tfrac{1}{2}(0.012\ \text{kg})v^2,$$

from which we find the maximum speed of the mass to be 1.63 m/s.

The oscillator has a wide variety of applications throughout theoretical physics for at least two reasons:

- Solutions to the equations of motion (both classical and quantum) can be expressed in terms of well-known mathematical functions.
- For small displacements from equilibrium, many real physical systems exhibit behavior approximately like that of a simple harmonic oscillator (e.g., a pendulum, an atom in a molecule, a neutron in an atomic nucleus).

Example 6.

An atom in a sodium crystal moves with small oscillations about its equilibrium position. The time required for one complete oscillation is about 3.10×10^{-13} s. The maximum displacement from equilibrium is 2.00×10^{-11} m. Approximate the system as a one-dimensional classical harmonic oscillator and estimate the maximum speed and the total energy of this atom. The mass of a sodium atom is about 3.80×10^{-26} kg.

Solution.

First, using Eq. (1.11) to calculate the force constant k from the given values of T and m, we get $k = 15.61$ N/m. With this value of k and the given amplitude, Eq. (1.9) yields for the maximum potential energy of the atom, $PE_{max} = 3.12 \times 10^{-21}$ J. Since the kinetic energy is zero when the potential energy is maximum, Eq. (1.10) shows that this is also the total energy of the oscillating atom. Again, from Eq. (1.10) we see that when the kinetic energy is maximum, the elastic potential energy is zero. Hence, the maximum kinetic energy is

$$KE_{max} = \tfrac{1}{2}mv_{max}^2 = 3.12 \times 10^{-21} J = \tfrac{1}{2}(3.80 \times 10^{-26} \text{ kg})v_{max}^2,$$

from which we obtain a maximum speed of $v_{max} = 405$ m/s.

1.5 A Rigid Rotator

The model of a rotating body has many applications in both classical and quantum physics (e.g., a spinning top, a gyro, a molecule, an atomic nucleus). The angular momentum of an extended body (not a *point* mass) rotating about a principal axis of rotation is given by

$$L = I\omega,$$

where ω is the magnitude of the angular velocity of the rotation about this axis. The quantity I is an inertial parameter that measures the body's intrinsic tendency to resist change in its angular motion, hence the term *moment of inertia*. The moment of inertia I depends not only on the inertia (mass) of the body, but also on how the mass is distributed in the body. The moment of inertia relative to a principal axis of rotation is given by

$$I = \int_{body} s^2 dm, \tag{1.15}$$

where s is the perpendicular distance from the infinitesimal mass element dm to the axis of rotation.

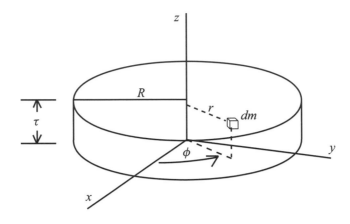

FIGURE 1.4. Moment of inertia of a uniform disk.

Example 7.

A thin, uniform disk of mass M has a radius R. Calculate the moment of inertia of this disk relative to an axis perpendicular to the plane of the disk and passing through its center.

Solution.

If τ represents the thickness of the disk, then the volume is $\pi R^2 \tau$. Choose the origin of the coordinate system at the center of the bottom of the disk with coordinates (r, ϕ, z) defined as shown in Fig. 1.4. The volume element occupied by the infinitesimal mass dm is $r\, dr\, d\phi\, dz$. From these definitions, we find that $s = r$ and $dm = (M/\pi R^2 \tau) r\, dr\, d\phi\, dz$. Substituting these expressions into Eq. (1.15), we obtain

$$I = \frac{M}{\pi R^2 \tau} \int_0^R r^3 dr \int_0^{2\pi} d\phi \int_0^\tau dz = \tfrac{1}{2} M R^2.$$

The moment of inertia about an axis passing through the center and perpendicular to the plane of a uniform disk of mass M and radius R is $\tfrac{1}{2} M R^2$.

1.6 Exercises

1. Two electric charges are located at the vertices of a right triangle, as illustrated in Fig. 1.5. Calculate the electric field (magnitude and direction) at the point P at the vacant vertex of the triangle.
2. Positive electric charges are located at three corners of a rectangle as shown in Fig. 1.6. Find the electric field (magnitude and direction) at the fourth corner.

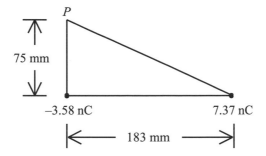

FIGURE 1.5. Charge distribution for Exercise 1.1.

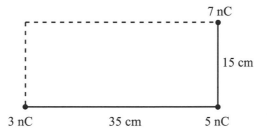

FIGURE 1.6. Charge distribution for Exercise 1.2.

3. A positive electric charge of 9 μC stands 13 mm to the left of a negative electric charge of -16 μC. Find the point at which the resultant electric field is zero.

4. A positive (8-μC) electric charge is placed 11 cm to the left of a negative (-5-μC) charge. Find the magnitude and direction of the resultant electric field at a point that is 5 cm directly above the midpoint between the two charges.

5. A narrow ion beam of uniform cross section consists of identical atoms each with three electrons removed. The beam is incident perpendicular to the plane of a metal target. The beam spot on the target has a diameter of 2.1 mm. The electric current represented by this ion beam is 22 μA. The ions travel with an average speed 6.48×10^6 m/s. Calculate the average density (ions/cm^3) of ions in the beam.

6. Two copper electrodes at different electric potential are placed some distance apart in a solution containing Cu^{++} ions. In one minute, 2.6×10^{19} Cu^{++} ions are attracted to the lower-potential electrode. What is the electric current (in amperes) corresponding to this ion flow?

7. A beam of electrons is directed onto a metal foil. The average diameter of the beam is 1 mm and the average density of electrons in the beam is 4.2×10^6 electrons/cm^3. The average speed of the electrons is 1.9×10^7 m/s. Use these data to calculate the average beam current in amperes.

8. A small coil of copper wire consists of 85 identical, closely-wound square loops each of length 254 mm on a side. The coil lies in a plane perpendicular

to a uniform magnetic field of 2.6 T. When the magnetic field is switched off, an average emf of 1835 volts is induced in the coil. Calculate the length of the time interval over which the field collapses to zero.

9. A uniform magnetic field of 2.2 T is directed perpendicular to the plane of a 2 cm × 3 cm rectangular loop of wire. Suddenly, the field is switched off and drops uniformly to zero in 0.04 s. Calculate the average voltage induced in the loop.

10. A circular coil of diameter 13.0 cm consists of 750 turns of wire. The plane of the coil is perpendicular to the earth's magnetic field. The coil is quickly rotated (one-fourth of a full rotation) about a horizontal axis through the plane of the coil in 3.27 milliseconds (ms) so that the plane of the coil is parallel to the earth's magnetic field. This induces an average emf of 155 millivolts (mV) in the coil. Calculate the value of the earth's magnetic field at this location.

11. A rectangular loop of wire lies partially in a uniform magnetic field of 2.3 T directed perpendicular to the plane of the loop as illustrated in Fig. 1.7. Without changing its orientation, the loop is pulled out of the field at a constant velocity of 7 m/s. Calculate the emf induced in the loop while it is being removed from the field.

12. A 123-g object attached to an elastic spring executes simple harmonic motion along a straight horizontal track. When the object is 8.34 cm from its equilibrium position, its speed is 3.87 m/s. The object makes 10 complete oscillations each second. Calculate the total energy of the object and the amplitude of the oscillation. Neglect frictional effects.

13. A small object oscillates horizontally with simple harmonic motion about its equilibrium position. At one instant, the object is seen to be 51.3 cm to the right of its equilibrium position and moving to the right with a speed of 6.71 m/s. Later, it is observed 23.8 cm to the right of the equilibrium point and moving left at 9.28 m/s. The total energy of this oscillator is 7.15 J. Neglect frictional effects and calculate the mass of this object.

14. A small object of mass 87 g is constrained to move horizontally in one dimension under the influence of a force that obeys Hooke's law. The period

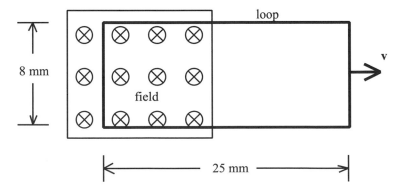

FIGURE 1.7. Rectangular loop in a magnetic field for Exercise 1.11.

of oscillation is 75 ms. At one instant, the object is located 3.73 cm to the left of its equilibrium position and is moving to the right with speed 5.29 m/s. Calculate the total energy of this oscillating object. What is the amplitude of the oscillation? Neglect friction.

15. A stable atomic nucleus is held together by the mutual attraction of the constituent particles (neutrons and protons) interacting through the nuclear force. In a rough approximation, each particle in the nucleus can be considered to be bound to the system by the combined effect of all the other particles. Assume that this average effect can be approximated by a one-dimensional classical harmonic oscillator. The maximum speed of a neutron bound in a certain spherical nucleus of radius 6.40×10^{-15} m is 7.35×10^6 m/s. The mass of a neutron is about 1.67×10^{-27} kg. Use these data to estimate the maximum classical binding force experienced by this neutron. (Of course, an atomic nucleus is *not* a classical system and quantum mechanics is required for a correct description.)

16. For a particle moving in one dimension along the x-axis, the velocity v and acceleration a are

$$v = \frac{dx}{dt} \qquad \text{and} \qquad a = \frac{dv}{dt},$$

respectively. From these definitions, show that the expressions for velocity and acceleration given in Eqs. (1.13) and (1.14) follow from Eq. (1.12).

17. A small mass attached to an elastic spring is constrained to move horizontally in one dimension. The mass is displaced 9 centimeters to the right of its equilibrium position and released. In 20 milliseconds it moves 3 cm to the left of the point from which it was released. Calculate the period of this oscillator. Neglect frictional effects.

18. A small mass released from rest at time $t = 0$ is observed to move with simple harmonic motion making 17 complete oscillations in 4 seconds. At the time 28.4 ms after its release, the mass is 53 mm from its equilibrium position and moving with speed 1.34 m/s. Calculate the amplitude of the oscillation and the maximum speed attained by the mass during its motion.

19. An object connected to an elastic spring rests on a horizontal, frictionless surface. The object is pulled away from its equilibrium position with a maximum force of 15.9 N and released from rest at time $t = 0$. The speed of the object increases and 32 ms after release it is moving at 3.88 m/s. The maximum speed reached by the object during its motion is 5.34 m/s. Calculate the amplitude and period of oscillation of this object. How far does the object move in the 32 ms following its release? What is the mass of this object?

20. A uniform metal rod rotates in a horizontal plane about a vertical axis that passes through its center of mass, as illustrated in Fig. 1.8. The rod is 40 cm long, has a mass of 200 g, and makes 1000 complete rotations each minute. Calculate the angular momentum of the rod about this axis.

21. Calculate the moment of inertia about an axis through the center of mass of the 3-kg rectangular block illustrated in Fig. 1.9.

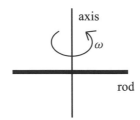

FIGURE 1.8. Uniform metal rod for Exercise 1.20.

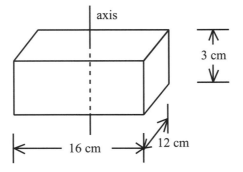

FIGURE 1.9. Uniform rectangular block for Exercise 1.21.

22. A thin, circular disk of mass M has a radius R. The density of the disk increases linearly with distance r from the z-axis passing through the center of the disk. (See Fig. 1.4.) In cylindrical coordinates, an infinitesimal element of mass is given by

$$dm = \frac{3M}{2\pi R^3 \tau} r^2 \, dr \, d\phi \, dz,$$

where τ represents the thickness of the disk. Show that integrating this expression over the volume of the disk yields the mass M. Calculate the moment of inertia of the disk relative to the z-axis.

2

Vectors

When describing the behavior of a physical system, we specify certain physical quantities at various points in space and time. For example, a particle of a given *mass* may be described by its *position* and *velocity* at each space-time point. The behavior of air in a room can depend on the *temperature* at each point in the room. Some of these quantities—*mass, temperature, density*—are completely specified by a single number with the appropriate units i.e., by a *magnitude* only. Such a physical quantity is called a *scalar* and is represented by a mathematical function of space and time called a *scalar point function*. We shall use italic type for symbols representing scalars (e.g., T for temperature). Others of these quantities (e.g., *velocity, acceleration, force*) require not only a magnitude but also a *direction* in order to be completely specified. For example, to specify the velocity of a particle, we must not only tell how fast the particle is moving (magnitude) but also tell in what direction it is traveling. A physical quantity of this kind is called a *vector* and is represented by a mathematical function of space and time called a *vector point function*. To denote a vector, we shall use boldface roman type (e.g., **F** for force).

2.1   Representations of Vectors

Pictorially, we represent a vector as a directed line segment—a straight line with an arrowhead on one end. The length of the line corresponds to the magnitude of the vector and the arrowhead indicates its direction. Two vectors, **a** and **b**, are

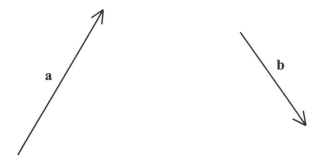

FIGURE 2.1. Graphical representation of vectors.

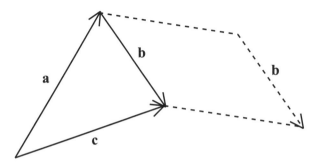

FIGURE 2.2. Addition of two vectors.

shown in Fig. 2.1. We assume that the same physical quantity (e.g., force) is represented by both. The magnitude of **a** is one and one-half times as large as the magnitude of **b**.

The addition of two vectors is defined in the following way: We add **b** to **a** by translating **b** parallel to itself (as indicated by the dashed lines in Fig. 2.2) until the tail of **b** coincides with the head of **a**. Note that this operation does not change **b** since its length (magnitude) and direction are just the same as before the operation. The resultant vector **c** given by

$$\mathbf{c} = \mathbf{a} + \mathbf{b}$$

is obtained by constructing a line from the tail of **a** to the head of **b** as indicated in Fig. 2.2.

We can extend this definition to include the difference **d** between two vectors,

$$\mathbf{d} = \mathbf{a} - \mathbf{b},$$

by reversing the direction of **b** (put the arrowhead on the other end) to obtain $-\mathbf{b}$ and then following the above rule for addition. Thus,

$$\mathbf{d} = \mathbf{a} + (-\mathbf{b}) = \mathbf{a} - \mathbf{b}.$$

This is illustrated in Fig. 2.3.

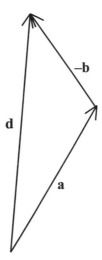

FIGURE 2.3. Difference of two vectors.

A vector can be represented in terms of its projections onto other vectors. For example, we can set up a rectangular coordinate system with origin at the tail of a vector **a** as shown in Fig. 2.4. We define three **unit** vectors (i.e., vectors with magnitudes equal to one), **i**, **j**, **k**, which are parallel to the x, y, and z axes, respectively. The vector **a** can then be written as

$$\mathbf{a} = \mathbf{i}\,a_x + \mathbf{j}\,a_y + \mathbf{k}\,a_z,$$

where a_x, a_y, and a_z are the projections of **a** onto the corresponding axes. (See Fig. 2.4.) The vector **a** can also be represented as an ordered triple,

$$\mathbf{a} = (a_x, a_y, a_z).$$

To find the magnitude of **a** in terms of its components, we first project **a** onto the xy-plane. The length of this projection (denoted by s in Fig. 2.4) is given by the Pythagorean theorem,

$$s = \sqrt{a_x^2 + a_y^2}.$$

Since the z-axis is perpendicular to the line s lying in the xy-plane, a further application of the Pythagorean theorem yields for the magnitude of **a**:

$$a = \sqrt{s^2 + a_z^2} = \sqrt{a_x^2 + a_y^2 + a_z^2}.$$

We also introduce the notation $|\mathbf{a}|$ to denote the magnitude of a vector. Thus, for the vector **a** the magnitude is

$$|\mathbf{a}| = a = \sqrt{a_x^2 + a_y^2 + a_z^2}.$$

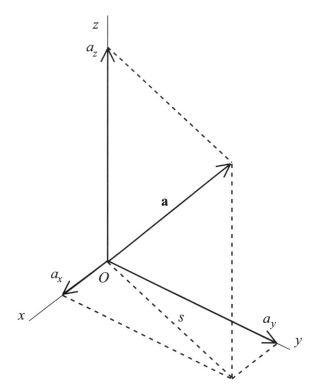

FIGURE 2.4. Rectangular components of a vector.

For example, we shall represent the position of a particle relative to a point O (see Fig. 2.4) by the vector

$$\mathbf{r} = \mathbf{i}\,x + \mathbf{j}\,y + \mathbf{k}\,z = (x, y, z), \tag{2.1}$$

with magnitude

$$|\mathbf{r}| = r = \sqrt{x^2 + y^2 + z^2}.$$

2.2   The Scalar Product of Two Vectors

Now, let us consider products of vectors. We define the *dot* product (the dot "·" is the symbol for the product operator) of two vectors $\mathbf{a}$ and $\mathbf{b}$ as

$$\mathbf{a} \cdot \mathbf{b} = a_x b_x + a_y b_y + a_z b_z.$$

Clearly, the dot product is commutative. That is,

$$\mathbf{a} \cdot \mathbf{b} = \mathbf{b} \cdot \mathbf{a}.$$

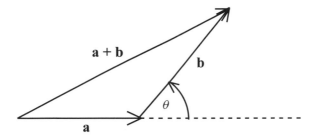

FIGURE 2.5. Angle between two vectors.

Next, we show that $\mathbf{a} \cdot \mathbf{b}$ is independent of the coordinate system we use. Note that

$$\begin{aligned}
|\mathbf{a} + \mathbf{b}|^2 &= (a_x + b_x)^2 + (a_y + b_y)^2 + (a_z + b_z)^2 \\
&= a^2 + b^2 + 2a_x b_x + 2a_y b_y + 2a_z b_z \\
&= a^2 + b^2 + 2\mathbf{a} \cdot \mathbf{b}.
\end{aligned} \tag{2.2}$$

By applying the law of cosines in Fig. 2.5, we also find that

$$\begin{aligned}
|\mathbf{a} + \mathbf{b}|^2 &= |\mathbf{a}|^2 + |\mathbf{b}|^2 - 2|\mathbf{a}||\mathbf{b}| \cos(\pi - \theta) \\
&= a^2 + b^2 + 2ab \cos\theta.
\end{aligned} \tag{2.3}$$

On comparing Eqs. (2.2) and (2.3), we see that

$$\mathbf{a} \cdot \mathbf{b} = ab \cos\theta, \tag{2.4}$$

which shows that $\mathbf{a} \cdot \mathbf{b}$ depends only on the magnitudes of $\mathbf{a}$ and $\mathbf{b}$ and the angle between them, independent of the coordinate system in which they may be represented. Clearly, if $\mathbf{a} \neq 0$, $\mathbf{b} \neq 0$, and $\mathbf{a} \cdot \mathbf{b} = 0$, then $\mathbf{a}$ and $\mathbf{b}$ must be mutually perpendicular.

We see that the dot product of two vectors is itself a scalar. For this reason, $\mathbf{a} \cdot \mathbf{b}$ is often called the *scalar* product of the vectors $\mathbf{a}$ and $\mathbf{b}$.

2.2.1   MECHANICAL WORK

Consider the displacement of a particle of mass m under the influence of a force $\mathbf{F}$ as illustrated in Fig. 2.6. In the case where some component of the force acts along the direction of the displacement, we say that *work* is done. To make this notion quantitative, we define the work ΔW done by the force $\mathbf{F}$ in displacing the particle by $\Delta\mathbf{s}$ to be equal to the magnitude of the displacement multiplied by the component of the force along the displacement. That is,

$$\Delta W = (F \cos\theta)\Delta s,$$

where θ is the angle between the force and displacement vectors. Consistent with this definition, we can use the scalar product to obtain a more compact expression

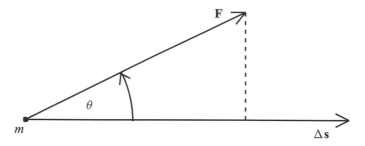

FIGURE 2.6. Displacement of a particle by a force.

for the work

$$\Delta W = \mathbf{F} \cdot \Delta \mathbf{s},$$

or for an infinitesimal displacement $d\mathbf{s}$,

$$dW = \mathbf{F} \cdot d\mathbf{s}. \tag{2.5}$$

2.3   The Vector Product of Two Vectors

It is also useful to define a different kind of product of two vectors called the *cross* product (the cross "$\times$" is the operator in this case),

$$\mathbf{a} \times \mathbf{b} = (a_y b_z - a_z b_y)\mathbf{i} + (a_z b_x - a_x b_z)\mathbf{j} + (a_x b_y - a_y b_x)\mathbf{k}. \tag{2.6}$$

Because this quantity is a vector, it is sometimes referred to as the *vector* product of $\mathbf{a}$ and $\mathbf{b}$. We also see that the cross product is *not* commutative, since

$$\mathbf{a} \times \mathbf{b} = -\mathbf{b} \times \mathbf{a}.$$

Clearly,

$$\mathbf{a} \times \mathbf{a} = \mathbf{b} \times \mathbf{b} = 0. \tag{2.7}$$

Assume that two vectors $\mathbf{a}$ and $\mathbf{b}$ are not parallel (or antiparallel). Then, from the rules defining the dot and cross products, we find that

$$\mathbf{a} \cdot (\mathbf{a} \times \mathbf{b}) = a_x(a_y b_z - a_z b_y) + a_y(a_z b_x - a_x b_z) + a_z(a_x b_y - a_y b_x) = 0.$$

Similarly,

$$\mathbf{b} \cdot (\mathbf{a} \times \mathbf{b}) = 0.$$

Thus, $\mathbf{a} \times \mathbf{b}$ is perpendicular to both $\mathbf{a}$ and $\mathbf{b}$. That is, $\mathbf{a} \times \mathbf{b}$ is perpendicular to the plane defined by $\mathbf{a}$ and $\mathbf{b}$ and in the direction in which a right-handed screw advances if $\mathbf{a}$ is turned into $\mathbf{b}$, as illustrated in Fig. 2.7. This determines the *direction* of $\mathbf{a} \times \mathbf{b}$ independent of the coordinate system.

To check this, we take $\mathbf{a} = \mathbf{i}$ and $\mathbf{b} = \mathbf{j}$ and use Eq. (2.6) to obtain

$$\mathbf{a} \times \mathbf{b} = \mathbf{i} \times \mathbf{j} = \mathbf{k}.$$

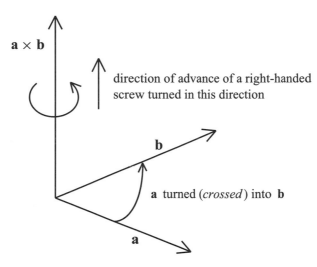

FIGURE 2.7. The cross product of two vectors.

Clearly, $\mathbf{i} \times \mathbf{j}$ is along the positive z-axis, which is consistent with the right-handed coordinate system we have used. Following the same procedure, we get

$$\mathbf{b} \times \mathbf{a} = \mathbf{j} \times \mathbf{i} = -\mathbf{k}.$$

So, $\mathbf{j} \times \mathbf{i}$ is directed along the *negative* z-axis.

Now let's consider the magnitude of $\mathbf{a} \times \mathbf{b}$,

$$|\mathbf{a} \times \mathbf{b}|^2 = (a_y b_z - a_z b_y)^2 + (a_z b_x - a_x b_z)^2 + (a_x b_y - a_y b_x)^2.$$

On expanding and collecting terms, we obtain

$$|\mathbf{a} \times \mathbf{b}|^2 = a^2 b^2 - a_x^2 b_x^2 - a_y^2 b_y^2 - a_z^2 b_z^2 - 2(a_y b_y a_z b_z + a_x b_x a_z b_z + a_x b_x a_y b_y).$$

Again collecting terms, this reduces to

$$|\mathbf{a} \times \mathbf{b}|^2 = a^2 b^2 - (\mathbf{a} \cdot \mathbf{b})(\mathbf{a} \cdot \mathbf{b})$$
$$= a^2 b^2 (1 - \cos^2 \theta)$$
$$= a^2 b^2 \sin^2 \theta$$

Thus,

$$|\mathbf{a} \times \mathbf{b}| = ab \sin \theta. \tag{2.8}$$

Hence, the cross product of two vectors $\mathbf{a}$ and $\mathbf{b}$ is independent of the coordinate system in which $\mathbf{a}$ and $\mathbf{b}$ are represented. From Eq. (2.8), it is clear that if $\mathbf{a}$ and $\mathbf{b}$ are parallel (or antiparallel), then $\mathbf{a} \times \mathbf{b} = 0$.

2.3.1   A CHARGED PARTICLE IN A MAGNETIC FIELD

To illustrate the usefulness of the cross product in physics, consider a small particle with electric charge q moving with velocity $\mathbf{v}$ in a magnetic field $\mathbf{B}$. From

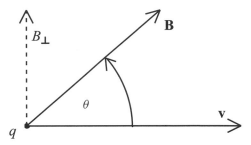

FIGURE 2.8. An electric charge q moving in a magnetic field.

experiment, one finds the magnitude of the magnetic force on the particle to be given by

$$F = qvB_\perp,$$

where $B_\perp$ is the component of $\mathbf{B}$ perpendicular to the velocity $\mathbf{v}$. The arrangement is illustrated in Fig. 2.8 from which it is clear that $B_\perp = B \sin \theta$ where θ is the angle between the vectors $\mathbf{v}$ and $\mathbf{B}$. Thus, the magnitude of the force is

$$F = qvB \sin \theta.$$

The direction of the magnetic force on the particle is observed in experiment to be perpendicular to the plane defined by $\mathbf{v}$ and $\mathbf{B}$. Hence, consistent with experiment, the magnitude and direction of the force on a charged particle moving in a magnetic field can be represented by the single expression

$$\mathbf{F} = q\mathbf{v} \times \mathbf{B}.$$

2.3.2   THE TRIPLE CROSS PRODUCT RULE

Next, we calculate the product,

$$\begin{aligned}
\mathbf{a} \times (\mathbf{b} \times \mathbf{c}) = {} & \mathbf{i}\,[a_y(b_x c_y - b_y c_x) - a_z(b_z c_x - b_x c_z)] \\
& + \mathbf{j}\,[a_z(b_y c_z - b_z c_y) - a_x(b_x c_y - b_y c_x)] \\
& + \mathbf{k}\,[a_x(b_z c_x - b_x c_z) - a_y(b_y c_z - b_z c_y)].
\end{aligned}$$

By adding and subtracting $a_x b_x c_x$ in the coefficient of $\mathbf{i}$ with similar terms added and subtracted in the coefficients of $\mathbf{j}$ and $\mathbf{k}$, we find that,

$$\begin{aligned}
\mathbf{a} \times (\mathbf{b} \times \mathbf{c}) &= \mathbf{i}[\mathbf{a} \cdot \mathbf{c}\, b_x - \mathbf{a} \cdot \mathbf{b}\, c_x] + \mathbf{j}[\mathbf{a} \cdot \mathbf{c}\, b_y - \mathbf{a} \cdot \mathbf{b}\, c_y] + \mathbf{k}[\mathbf{a} \cdot \mathbf{c}\, b_z - \mathbf{a} \cdot \mathbf{b}\, c_z] \\
&= (\mathbf{a} \cdot \mathbf{c})\mathbf{b} - (\mathbf{a} \cdot \mathbf{b})\mathbf{c}.
\end{aligned}$$

This relation is called the *triple cross product rule*,

$$\mathbf{a} \times (\mathbf{b} \times \mathbf{c}) = (\mathbf{a} \cdot \mathbf{c})\,\mathbf{b} - (\mathbf{a} \cdot \mathbf{b})\,\mathbf{c}. \tag{2.9}$$

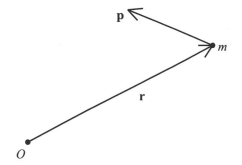

FIGURE 2.9. Force on particle of mass m at position $\mathbf{r}$.

2.3.3 Torque and Angular Momentum

For a particle of mass m, Newton's second law may be written as

$$\mathbf{F} = \frac{d\mathbf{p}}{dt}, \tag{2.10}$$

where the linear momentum $\mathbf{p}$ is related to the velocity $\mathbf{v}$ of the particle by $\mathbf{p} = m\mathbf{v}$. The *moment of the force* or *torque* τ is defined by

$$\tau = \mathbf{r} \times \mathbf{F}, \tag{2.11}$$

where the vector $\mathbf{r}$ represents the position of the particle relative to an arbitrary reference point O as illustrated in Fig. 2.9.

From Eqs. (2.10) and (2.11), we see that

$$\tau = \mathbf{r} \times \mathbf{F} = \mathbf{r} \times \left(\frac{d\mathbf{p}}{dt}\right) = \frac{d}{dt}(\mathbf{r} \times \mathbf{p}) - \frac{d\mathbf{r}}{dt} \times \mathbf{p}$$

$$= \frac{d}{dt}(\mathbf{r} \times \mathbf{p}) - \frac{1}{m}\mathbf{p} \times \mathbf{p} = \frac{d}{dt}(\mathbf{r} \times \mathbf{p}), \tag{2.12}$$

where we have used Eq. (2.7) in the last step. The quantity $\mathbf{r} \times \mathbf{p}$ is the *moment of linear momentum* or *angular momentum* $\mathbf{L}$ of the particle relative to the point O. With these definitions of torque and angular momentum, we have from Eq. (2.12),

$$\tau = \frac{d\mathbf{L}}{dt}, \tag{2.13}$$

which is the rotational analogue of Eq. (2.10) describing the translational motion of the particle.

Example 1.
The position of a particle of mass m with respect to a point O is given by the vector $\mathbf{r}$ as shown in Fig. 2.10. The particle moves with constant angular velocity ω about an axis passing through O and making an angle θ with respect to $\mathbf{r}$. Obtain an expression for the particle's linear velocity $\mathbf{v}$ in terms of the given parameters. Calculate the angular momentum of this particle with respect to the point O.

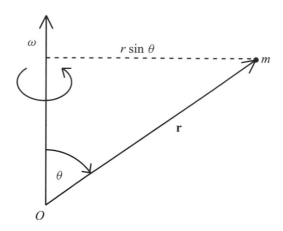

FIGURE 2.10. A particle moving relative to a point O.

Solution.

It is clear from Fig. 2.10 that the speed v of the particle is given by

$$v = \omega(r \sin \theta).$$

It is also clear from this drawing that the *direction* of the velocity is perpendicular to the plane defined by the vectors ω and $\mathbf{r}$. Thus, the vector with the required magnitude and direction for the linear velocity $\mathbf{v}$ is

$$\mathbf{v} = \omega \times \mathbf{r}. \tag{2.14}$$

The angular momentum of the particle is

$$\mathbf{L} = \mathbf{r} \times \mathbf{p} = \mathbf{r} \times (m\mathbf{v}) = m[\mathbf{r} \times (\omega \times \mathbf{r})] = m[r^2\omega - (\mathbf{r} \cdot \omega)\mathbf{r}]. \tag{2.15}$$

Note that the direction of the vector $\mathbf{r}$ changes with time t. Choosing the z-axis along the direction of ω, we write $\mathbf{r}$ in terms of its rectangular components,

$$\mathbf{r} = \mathbf{i}\,(r \sin \theta \cos(\phi + \omega t)) + \mathbf{j}\,(r \sin \theta \sin(\phi + \omega t)) + \mathbf{k}\,(r \cos \theta).$$

In representing $\mathbf{r}$ in spherical polar coordinates in this example, r and θ are constant and ϕ is the azimuth angle at time $t = 0$. Thus, for the angular momentum of the particle, we have

$$\mathbf{L} = m\omega r^2[-\mathbf{i} \sin \theta \cos \theta \cos(\phi + \omega t) - \mathbf{j} \sin \theta \cos \theta \sin(\phi + \omega t) + \mathbf{k} \sin^2 \theta],$$

which shows the time dependence explicitly. In arriving at this last result, we have used $\mathbf{r} \cdot \omega = r\omega \cos \theta$ in Eq. (2.15).

The time T required for one complete revolution of the particle is $T = 2\pi/\omega$. The angular momentum averaged over one revolution is

$$\bar{\mathbf{L}} = \frac{1}{T} \int_0^T \mathbf{L}\,dt = \omega m r^2 \sin^2 \theta,$$

which is directed along the rotation axis, as we expect from the symmetry in this example.

Example 2.

Use the result in Example 1 to calculate the angular momentum of a rigid sphere of radius R, mass M, and uniform density $\rho = 3M/4\pi R^3$ rotating with constant angular velocity $\boldsymbol{\omega}$ about an axis passing through the geometrical center of the sphere.

Solution.

Let $d\mathbf{L}$ represent the angular momentum of an infinitesimal mass element $dm = \rho d^3r = \rho r^2 \sin\theta \, d\phi \, d\theta \, dr$. The total angular momentum of the sphere is obtained by summing these contributions. From the result in Example 1, it follows that the angular momentum of the sphere at time t is

$$\mathbf{L} = \int_{\text{sphere}} d\mathbf{L} = \frac{3M\omega}{4\pi R^3} \int_0^R \int_0^\pi \int_0^{2\pi} \left\{ -\mathbf{i}\left[\sin^2\theta\cos\theta\cos(\phi+\omega t)\right] \right.$$
$$-\mathbf{j}\left[\sin^2\theta\cos\theta\sin(\phi+\omega t)\right]$$
$$\left. +\mathbf{k}\left[\sin^3\theta\right]\right\} r^4 \, d\phi \, d\theta \, dr.$$

The time dependence is illusory, because

$$\int_0^{2\pi} \cos(\phi+\omega t)\,d\phi = \int_0^{2\pi} \sin(\phi+\omega t)\,d\phi = 0.$$

Thus, the components of $\mathbf{L}$ perpendicular to the rotation axis vanish and the angular momentum of the spinning sphere is

$$\mathbf{L} = \mathbf{k}\frac{3M\omega}{4\pi R^3} \int_0^R r^4\,dr \int_0^\pi \sin^3\theta\,d\theta \int_0^{2\pi} d\phi = \frac{2}{5}MR^2\boldsymbol{\omega}.$$

Note that the angular momentum of this uniformly rotating sphere is independent of time and lies in the same direction as the rotation axis, as is to be expected from the symmetry.

2.3.4   SCALAR PRODUCT WITH CROSS PRODUCT

Using rectangular components, we expand the scalar product of a vector $\mathbf{a}$ with the cross product of two other vectors $\mathbf{b}$ and $\mathbf{c}$,

$$\mathbf{a}\cdot(\mathbf{b}\times\mathbf{c}) = a_x(b_yc_z - b_zc_y) + a_y(b_zc_x - b_xc_z) + a_z(b_xc_y - b_yc_x)$$
$$= (a_yb_z - a_zb_y)c_x + (a_zb_x - a_xb_z)c_y + (a_xb_y - a_yb_x)c_z$$
$$= (\mathbf{a}\times\mathbf{b})\cdot\mathbf{c}.$$

Clearly, this result is independent of the coordinate system in which the vectors are represented and we can make the general statement: *In the scalar product of a vector with the cross product of two other vectors, the dot (·) and cross (×) operators may be interchanged without affecting the result.* That is,

$$\mathbf{a}\cdot(\mathbf{b}\times\mathbf{c}) = (\mathbf{a}\times\mathbf{b})\cdot\mathbf{c}. \tag{2.16}$$

2.4   Exercises

1. Relative to a single cartesian coordinate system, the components of four
   vectors **A**, **B**, **C**, and **D** are given by

$$\mathbf{A} = 3\mathbf{i} - \mathbf{j} + 2\mathbf{k} \qquad\qquad \mathbf{C} = -2\mathbf{i} - 2\mathbf{j} - 6\mathbf{k}$$
$$\mathbf{B} = -\mathbf{i} + 5\mathbf{j} + 4\mathbf{k} \qquad\qquad \mathbf{D} = 6\mathbf{i} + 2\mathbf{j} + 11\mathbf{k}$$

   Express the following vectors **R**, **S**, and **T** in terms of their cartesian
   coordinates,

$$\mathbf{R} = \mathbf{A} + \mathbf{B} + \mathbf{C}$$
$$\mathbf{S} = 2\mathbf{A} - 3\mathbf{C}$$
$$\mathbf{T} = \mathbf{A} \times (\mathbf{B} \times \mathbf{C}).$$

   Examine carefully each of the original vectors, **A**, **B**, **C**, and **D** and the three
   vectors, **R**, **S**, and **T**. Make a statement (*different for each*) about the direction
   of each of the three **R**, **S**, and **T**. Which of the three statements could still be
   made if the vectors were expressed in terms of any other coordinate system?
   Explain. Which of the three statements would *not* be applicable, if the vectors
   were related to some other arbitrary coordinate system? Explain.

2. A hiker makes the following successive displacements: 9.18 km, 33° N of E;
   7.73 km, 69° S of E; 5.22 km, 16° S of W. Find the resultant displacement of
   the hiker.

3. The rectangular components of two vectors **a** and **b** are given by

$$\mathbf{a} = 3\mathbf{i} - 6\mathbf{j} + 9\mathbf{k} \qquad\qquad \mathbf{b} = 6\mathbf{i} + 2\mathbf{j} - 5\mathbf{k}.$$

   Calculate the magnitude of each of these two vectors. Calculate the sum of **a**
   and **b**; the difference; the scalar (dot) product; the vector (cross) product.

4. Three vectors, **u**, **v**, and **w** are given by

$$\mathbf{u} = -4\mathbf{i} + 3\mathbf{j} + 7\mathbf{k} \qquad \mathbf{v} = 8\mathbf{i} - 5\mathbf{j} - 3\mathbf{k} \qquad \mathbf{w} = \mathbf{i} + 6\mathbf{j} - 2\mathbf{k}.$$

   Calculate the following:

   a. $2\mathbf{u} - 3\mathbf{v} + 6\mathbf{w}$
   b. $\mathbf{u} \cdot (\mathbf{v} \times \mathbf{w})$
   c. $\mathbf{v} \cdot (\mathbf{u} \times \mathbf{w})$
   d. $(\mathbf{u} \times \mathbf{v}) \cdot \mathbf{w}$
   e. $\mathbf{u} \times (\mathbf{v} \times \mathbf{w})$
   f. $(\mathbf{u} \times \mathbf{v}) \times \mathbf{w}$

5. Given the following vectors

$$\mathbf{p} = (-3, 7, 8) \qquad \mathbf{q} = (6, -4, 5) \qquad \mathbf{r} = (5, 2, -6)$$

   calculate

   a. $3\mathbf{p} - 2\mathbf{q} + 5\mathbf{r}$
   b. $\mathbf{p} + \mathbf{q} - \mathbf{r}$

c. $\mathbf{p} \times \mathbf{q} - \mathbf{r}$

d. $\mathbf{p} \cdot \mathbf{q} - \mathbf{r} \cdot \mathbf{p} + \mathbf{q} \cdot \mathbf{r}$

e. $(\mathbf{p} - 3\mathbf{q}) \times \mathbf{r}$

6. Use the scalar product to find the rectangular components of a vector $\mathbf{c}$ which is perpendicular to the two vectors $\mathbf{a}$ and $\mathbf{b}$ given by

$$\mathbf{a} = (-2, 6, 5) \qquad\qquad \mathbf{b} = (3, -1, -4).$$

Now use the *vector* product to find the rectangular components of a vector perpendicular to $\mathbf{a}$ and $\mathbf{b}$. Do your results from these two methods agree? Explain.

7. Repeat Exercise 2.6 for each of the following pairs of vectors.

$$\mathbf{a} = (2, -5, -3) \qquad\qquad \mathbf{b} = (-1, 7, 4)$$
$$\mathbf{a} = (4, -1, -6) \qquad\qquad \mathbf{b} = (8, -4, -15)$$
$$\mathbf{a} = (1, -3, 2) \qquad\qquad \mathbf{b} = (-4, 7, -5)$$
$$\mathbf{a} = (3, 10, 4) \qquad\qquad \mathbf{b} = (1, 8, 4)$$

8. Let $\mathbf{a}$ and $\mathbf{b}$ be two arbitrary nonzero vectors that are neither parallel, anti-parallel, nor perpendicular to each other. A third arbitrary vector $\mathbf{c}$ may be expressed as a linear combination of $\mathbf{a}$ and $\mathbf{b}$,

$$\mathbf{c} = \alpha \mathbf{a} + \beta \mathbf{b}.$$

Obtain expressions for the coefficients α and β explicitly in terms of the three vectors $\mathbf{a}$, $\mathbf{b}$, and $\mathbf{c}$.

9. If the position of a particle is represented by the vector $\mathbf{r}$, the velocity $\mathbf{v}$ and acceleration $\mathbf{a}$ of the particle are given by the time derivatives

$$\mathbf{v} = \frac{d\mathbf{r}}{dt} \qquad \text{and} \qquad \mathbf{a} = \frac{d\mathbf{v}}{dt}.$$

Show that

$$\frac{d}{dt}[\mathbf{r} \times (\mathbf{v} \times \mathbf{r})] = r^2 \mathbf{a} + (\mathbf{r} \cdot \mathbf{v})\mathbf{v} - (v^2 + \mathbf{r} \cdot \mathbf{a})\mathbf{r}.$$

10. Given the four arbitrary vectors $\mathbf{a}$, $\mathbf{b}$, $\mathbf{p}$, and $\mathbf{q}$, use the result in Eq. (2.16) to show that

$$(\mathbf{a} \times \mathbf{b}) \cdot (\mathbf{p} \times \mathbf{q}) = (\mathbf{a} \cdot \mathbf{p})(\mathbf{b} \cdot \mathbf{q}) - (\mathbf{a} \cdot \mathbf{q})(\mathbf{b} \cdot \mathbf{p}).$$

11. Three vectors $\mathbf{s}$, $\mathbf{t}$, and $\mathbf{u}$ satisfy the condition

$$\mathbf{s} \times (\mathbf{t} \times \mathbf{u}) = 0.$$

Describe clearly and completely two *different* geometrical relationships among these vectors that could give rise to this condition. Be very clear.

12. Suppose $\mathbf{n}$ is a *unit* vector perpendicular to a given vector $\mathbf{b}$. Give as complete a description as possible (magnitude and direction) of the vector

$$\mathbf{n} \times (\mathbf{n} \times \mathbf{b}).$$

Write your answer in a complete sentence.

13. An arbitrary vector **u** can be expressed in terms of its components relative to a unit vector **n**. That is,

$$\mathbf{u} = \mathbf{u}_\parallel + \mathbf{u}_\perp,$$

where $\mathbf{u}_\parallel$ is the component of **u** parallel to **n** and $\mathbf{u}_\perp$ is the component of **u** perpendicular to **n**. Use the triple cross product rule in Eq. (2.9) to construct $\mathbf{u}_\parallel$ and $\mathbf{u}_\perp$ in terms of **n**.

14. Three vectors **a**, **b**, and **p** satisfy the relation

$$\mathbf{b} \times (\mathbf{a} \times \mathbf{p}) = \lambda \mathbf{a},$$

where λ is a scalar. Find the condition that the vectors **a** and **b** must satisfy. Draw a clearly labeled vector diagram showing explicitly the geometrical relationship between **a** and **b**. Express λ explicitly in terms of the given vectors.

15. Show that the scalar product of two vectors **a** and **b** cannot be larger than the product of the magnitudes of the vectors. That is,

$$ab \geq \mathbf{a} \cdot \mathbf{b}.$$

This relation is known as the *Schwarz inequality*.

16. A particle with positive electric charge e moves with velocity **v** in a plane perpendicular to a magnetic field **B** of fixed direction. Calculate the torque on the particle relative to an arbitrary point in the plane of motion. From your result, describe in words the direction of this torque relative to the direction of motion of the particle and the direction of the magnetic field. Explain clearly how your answer follows from your calculation.

17. The linear momentum **p** of a particle of mass m is given by

$$\mathbf{p} = \frac{L}{r}\mathbf{n},$$

where **n** is a fixed unit vector, L is a constant, and r is the distance of the particle from a fixed point O. Calculate the angular momentum **L** of the particle relative to the point O. Express your result in terms of L, **n**, and the unit vector $\hat{\mathbf{r}}$ which gives the direction of the position of the particle relative to O. Use Eq. (2.13) to calculate the corresponding torque. From this result, find the force **F** on the particle. Express all results in terms of the given parameters m, L, and r and the unit vectors $\hat{\mathbf{r}}$ and **n**. Describe in words the behavior of the magnitude and direction of the force. Make a drawing showing the relative directions of the vectors **F**, $\hat{\mathbf{r}}$, and **n**. Now use Eq. (2.10) to obtain an expression for the force directly. Do your results agree?

18. Show that the magnitude of the force on the particle in Exercise 2.17 can be written as

$$F = \frac{L^2}{mb^3}\sin^3\theta\cos\theta,$$

where b is the distance of closest approach of the particle to the point O and θ is the angle between the unit vectors $\hat{\mathbf{r}}$ and **n**.

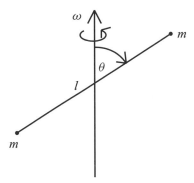

FIGURE 2.11. A rotating dumbbell for Exercise 2.19.

19. A dumbbell consists of two mass points of equal mass m at opposite ends of a rigid rod of length l and negligible mass. Calculate the angular momentum of this dumbbell as it rotates with constant angular velocity ω about an axis that makes a constant angle θ with the dumbbell and passes through its center of mass as illustrated in Fig. 2.11. Express your result so that the time dependence is explicit.

20. A sphere of mass M, radius R, and nonuniform density

$$\rho(\mathbf{r}) = \frac{15M}{8\pi R^3}\left(1 - \frac{r^2}{R^2}\right),$$

spins with a constant angular velocity ω about an axis passing through its center. Calculate the angular momentum of this spinning sphere.

21. Use the rule in Eq. (2.16) to obtain directly the result in Eq. (2.8).

22. Two vectors $\mathbf{a}$ and $\mathbf{b}$ are neither perpendicular, parallel, nor antiparallel to each other. Find a linear combination of $\mathbf{a}$ and $\mathbf{b}$ (call the combination $\mathbf{c}$) that is perpendicular to $\mathbf{a}$. Construct a third vector $\mathbf{d}$ that is perpendicular to $\mathbf{a}$ and $\mathbf{c}$. Express all results in terms of the original vectors $\mathbf{a}$ and $\mathbf{b}$.

23. Three vectors $\mathbf{a}, \mathbf{b}$, and $\mathbf{c}$ do not all lie in the same plane. None of these vectors is parallel (or antiparallel) to either of the other two and none is perpendicular to either of the other two. Verify that the vectors defined by

$$\mathbf{a}' = \mathbf{a},$$

$$\mathbf{b}' = \mathbf{b} - \frac{\mathbf{a}\cdot\mathbf{b}}{|\mathbf{a}|^2}\,\mathbf{a},$$

$$\mathbf{c}' = \mathbf{c} - \frac{[(\mathbf{a}\times\mathbf{b})\times\mathbf{c}]\times(\mathbf{a}\times\mathbf{b})}{|\mathbf{a}\times\mathbf{b}|^2},$$

are mutually perpendicular (*orthogonal*). From these, construct a set of three orthogonal *unit* vectors $\mathbf{n}_i$ (for $i = 1, 2, 3$) with the property

$$\mathbf{n}_i\cdot\mathbf{n}_j = \begin{cases} 1 & \text{for } i = j \\ 0 & \text{for } i \neq j. \end{cases}$$

Express your answer in terms of the original vectors $\mathbf{a}, \mathbf{b}$, and $\mathbf{c}$.

24. None of the four vectors **a**, **b**, **c** and **d** is parallel, antiparallel, or perpendicular to any of the other three and **c** does not lie in any plane defined by any pair of the other three. In a three-dimensional space, the four vectors cannot all be independent. Write **d** as a linear combination of **a**, **b**, and **c**. Express your result in terms of **a**, **b**, **c**, and **d**.

25. Given five vectors, **a**, **b**, **c**, **d**, and **e**, show that

$$[(\mathbf{a} \times \mathbf{b}) \times (\mathbf{c} \times \mathbf{d})] \cdot \mathbf{e} = \mathbf{a} \cdot [(\mathbf{c} \cdot \mathbf{e})(\mathbf{b} \times \mathbf{d}) - (\mathbf{d} \cdot \mathbf{e})(\mathbf{b} \times \mathbf{c})].$$

3
Vector Calculus

The physical quantities that we encounter in physics and applied mathematics (e.g., force, temperature, electric field) are represented by mathematical functions of space and time, hence, functions of several variables. For example, for the force $\mathbf{F}$ on a particle we may write

$$\mathbf{F} = \mathbf{F}(x, y, z, t),$$

which shows that the force depends explicitly on the three position coordinates (x, y, z) and the time coordinate t. The force on the particle depends on where the particle is located in space and on when it is at that point.

The equations that describe the behavior of matter and of radiation (e.g., Newton's second law, Maxwell's equations, the Schrödinger equation) involve space and time derivatives of the physical quantities. Because these quantities are functions of several variables, it is *partial* derivatives that become the focus of our interest here. For this reason, we begin our discussion of vector calculus with some salient features of partial differentiation and differentials of functions of several variables.

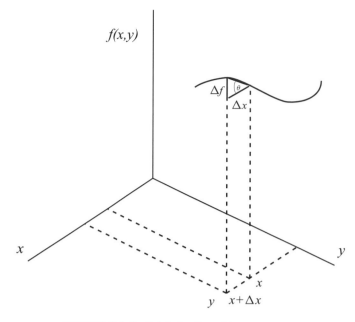

FIGURE 3.1. Partial derivative with respect to x.

3.1   Partial Derivatives[1]

Consider a continuous function f of two variables x and y,

$$f = f(x, y).$$

We define the *partial* derivative of f with respect to x by

$$\frac{\partial f}{\partial x} = \lim_{\Delta x \to 0} \frac{f(x + \Delta x, y) - f(x, y)}{\Delta x},$$

where y is held constant.

From Fig. 3.1, we see that for fixed y,

$$\tan \theta = \frac{\Delta f}{\Delta x} \xrightarrow{\Delta x \to 0} \frac{\partial f}{\partial x}.$$

Similarly, for the partial derivative of f with respect to y, we have

$$\frac{\partial f}{\partial y} = \lim_{\Delta y \to 0} \frac{f(x, y + \Delta y) - f(x, y)}{\Delta y},$$

where x remains constant.

---

[1]The discussion in this section follows that of W. Kaplan, *Advanced Calculus*, Addison-Wesley, Reading, MA, 1952, pp. 79-84 and M. L. Boas, *Mathematical Methods in the Physical Sciences*, J. Wiley and Sons, New York, 1983, pp. 150-153.

We define the *total* differential df by

$$df = \frac{\partial f}{\partial x} dx + \frac{\partial f}{\partial y} dy, \tag{3.1}$$

where $\partial f/\partial x$ and $\partial f/\partial y$ are continuous functions of x and y.

Let Δf denote the change in $f(x, y)$ when x and y are changed by small amounts to $x + \Delta x$ and $y + \Delta y$, respectively. That is,

$$\Delta f = f(x + \Delta x, y + \Delta y) - f(x, y).$$

We add and subtract $f(x + \Delta x, y)$ to get

$$\Delta f = f(x + \Delta x, y) - f(x, y) + f(x + \Delta x, y + \Delta y) - f(x + \Delta x, y).$$

According to the mean value theorem from calculus, we know that for a continuous function $g(x)$ whose derivative exists at all points between x and $x + \Delta x$, there is *some* point x' in this interval at which the line tangent to the curve $g(x)$ has the same slope as the line passing through the points on the curve at x and $x + \Delta x$ as shown in Fig. 3.2. That is,

$$\frac{\Delta g}{\Delta x} \equiv \frac{g(x + \Delta x) - g(x)}{\Delta x} = \left[\frac{dg}{dx}\right]_{x=x'}.$$

So, applying this theorem in Δf above we can write

$$\Delta f = \left(\frac{\partial f}{\partial x}\right)_{x',y} \Delta x + \left(\frac{\partial f}{\partial y}\right)_{x+\Delta x, y'} \Delta y,$$

where x' lies between x and $x + \Delta x$ and y' lies between y and $y + \Delta y$. The continuous partial derivatives in this equation differ from their values at the point (x, y) by small amounts α and β, respectively. Thus,

$$\Delta f = \left(\frac{\partial f}{\partial x} + \alpha\right) \Delta x + \left(\frac{\partial f}{\partial y} + \beta\right) \Delta y$$

$$= \left(\frac{\partial f}{\partial x} \Delta x + \frac{\partial f}{\partial y} \Delta y\right) + \alpha \Delta x + \beta \Delta y.$$

Since the partial derivatives are continuous, in the limit that Δx and Δy become infinitesimally small (in which case, we represent them by dx and dy, respectively), the quantities α and β also become infinitesimally small. Therefore, to first order in the infinitesimals, we have

$$\Delta f \longrightarrow \frac{\partial f}{\partial x} dx + \frac{\partial f}{\partial y} dy = df,$$

in accord with Eq. (3.1). This provides us with an intuitive understanding of the total differential. The definition in Eq. (3.1) is readily generalized to any number of independent variables $x_1, x_2, x_3, \ldots, x_n$.

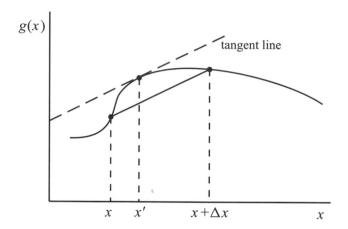

FIGURE 3.2. The mean value theorem.

3.2   A Vector Differential Operator

Maxwell's equations,[2] the ground on which classical electrodynamics is built, are coupled, first-order differential equations describing the electric and magnetic fields. These equations may be written very elegantly and compactly by introducing a vector differential operator denoted by ∇. (Similar applications arise in classical mechanics, quantum mechanics, and other branches of physics and applied mathematics.) We define this operator by

$$\nabla \equiv \mathbf{i}\frac{\partial}{\partial x} + \mathbf{j}\frac{\partial}{\partial y} + \mathbf{k}\frac{\partial}{\partial z}, \qquad (3.2)$$

where x, y, and z are the rectangular components of the position vector $\mathbf{r}$ of Eq. (2.1).

3.2.1   The Gradient

Let us apply this operator to a scalar function of position, $\psi(\mathbf{r})$,

$$\nabla\psi(\mathbf{r}) = \mathbf{i}\frac{\partial\psi}{\partial x} + \mathbf{j}\frac{\partial\psi}{\partial y} + \mathbf{k}\frac{\partial\psi}{\partial z}. \qquad (3.3)$$

Note that this operation on a *scalar* function yields a *vector* function which we call the *gradient of* ψ. It is sometimes denoted by **grad** ψ. This vector has components

$$\nabla\psi = \mathbf{i}(\nabla\psi)_x + \mathbf{j}(\nabla\psi)_y + \mathbf{k}(\nabla\psi)_z.$$

Clearly, the rectangular components of $\nabla\psi$ are

$$(\nabla\psi)_x = \frac{\partial\psi}{\partial x} \qquad (\nabla\psi)_y = \frac{\partial\psi}{\partial y} \qquad (\nabla\psi)_z = \frac{\partial\psi}{\partial z}.$$

---

[2]See Appendix B.

Example 1.

For the position vector $\mathbf{r} = \mathbf{i}\,x + \mathbf{j}\,y + \mathbf{k}\,z$, calculate the gradient of $1/r$ where $r = \sqrt{x^2 + y^2 + z^2}$ is the magnitude of $\mathbf{r}$.

Solution.

From the definition of the gradient in Eq. (3.3), we have

$$\nabla\left(\frac{1}{r}\right) = \left(\mathbf{i}\frac{\partial}{\partial x} + \mathbf{j}\frac{\partial}{\partial y} + \mathbf{k}\frac{\partial}{\partial z}\right)(x^2 + y^2 + z^2)^{-\frac{1}{2}}$$

$$= -(x^2 + y^2 + z^2)^{-\frac{3}{2}}(\mathbf{i}\,x + \mathbf{j}\,y + \mathbf{k}\,z) = -\frac{\mathbf{r}}{r^3}.$$

Thus,

$$\nabla\left(\frac{1}{r}\right) = -\frac{\mathbf{r}}{r^3}. \tag{3.4}$$

3.2.2 THE DIVERGENCE AND THE CURL

Now let us apply the vector differential operator to a *vector* function of position, $\mathbf{F}(\mathbf{r})$. There are two ways we can do this, using either the dot product rule or the cross product rule. These yield, respectively,[3]

$$\nabla \cdot \mathbf{F} = \frac{\partial F_x}{\partial x} + \frac{\partial F_y}{\partial y} + \frac{\partial F_z}{\partial z} \tag{3.5}$$

and

$$\nabla \times \mathbf{F} = \mathbf{i}\left(\frac{\partial F_z}{\partial y} - \frac{\partial F_y}{\partial z}\right) + \mathbf{j}\left(\frac{\partial F_x}{\partial z} - \frac{\partial F_z}{\partial x}\right) + \mathbf{k}\left(\frac{\partial F_y}{\partial x} - \frac{\partial F_x}{\partial y}\right). \tag{3.6}$$

The first of these is called the *divergence of* $\mathbf{F}$, which we denote by div $\mathbf{F}$. The second is known as the *curl of* $\mathbf{F}$, often written as **curl F**.

Consistent with the definitions of vector products, we can make successive applications of the differential operator ∇. For example, if we take $\mathbf{F}$ to be the gradient of some scalar function $\psi(\mathbf{r})$, then from the definition of the divergence, we see that

$$\nabla \cdot (\nabla\psi) = \frac{\partial^2\psi}{\partial x^2} + \frac{\partial^2\psi}{\partial y^2} + \frac{\partial^2\psi}{\partial z^2} \equiv \nabla^2\psi. \tag{3.7}$$

Here we have defined a differential operator by

$$\nabla^2 \equiv \frac{\partial^2}{\partial x^2} + \frac{\partial^2}{\partial y^2} + \frac{\partial^2}{\partial z^2}. \tag{3.8}$$

The quantity $\nabla^2\psi$ is called the *Laplacian of* ψ.

Note that the operator defined in Eq. (3.8) can be applied to each of the rectangular *components* of a vector $\mathbf{A}(\mathbf{r})$, (A_x, A_y, A_z), which are themselves *scalars*. Then,

---

[3]The applications are straightforward in rectangular coordinates, but more complicated in other coordinate systems, as we shall see.

multiplying by the appropriate unit vectors and adding the three terms together, we obtain

$$\nabla^2 \mathbf{A} = \frac{\partial^2 \mathbf{A}}{\partial x^2} + \frac{\partial^2 \mathbf{A}}{\partial y^2} + \frac{\partial^2 \mathbf{A}}{\partial z^2},$$

which is the *Laplacian of* **A**, a vector.

In Eq. (3.6), let $\mathbf{F} = \nabla \psi(\mathbf{r})$, where $\psi(\mathbf{r})$ is any scalar function of **r**. This gives

$$\nabla \times (\nabla \psi) = \mathbf{i}\left[\frac{\partial}{\partial y}\left(\frac{\partial \psi}{\partial z}\right) - \frac{\partial}{\partial z}\left(\frac{\partial \psi}{\partial y}\right)\right] + \mathbf{j}\left[\frac{\partial}{\partial z}\left(\frac{\partial \psi}{\partial x}\right) - \frac{\partial}{\partial x}\left(\frac{\partial \psi}{\partial z}\right)\right]$$

$$+ \mathbf{k}\left[\frac{\partial}{\partial x}\left(\frac{\partial \psi}{\partial y}\right) - \frac{\partial}{\partial y}\left(\frac{\partial \psi}{\partial x}\right)\right].$$

Clearly, the coefficient of each unit vector is zero. Thus, the curl of the gradient of *any scalar function* of **r** is identically zero,[4]

$$\nabla \times (\nabla \psi(\mathbf{r})) \equiv 0. \tag{3.9}$$

Now take **F** in Eq. (3.5) to be the curl of some other arbitrary vector, $\mathbf{A}(\mathbf{r})$. Then,

$$\nabla \cdot (\nabla \times \mathbf{A}(\mathbf{r})) = \frac{\partial}{\partial x}\left(\frac{\partial A_z}{\partial y} - \frac{\partial A_y}{\partial z}\right) + \frac{\partial}{\partial y}\left(\frac{\partial A_x}{\partial z} - \frac{\partial A_z}{\partial x}\right) + \frac{\partial}{\partial z}\left(\frac{\partial A_y}{\partial x} - \frac{\partial A_x}{\partial y}\right).$$

These second partial derivatives cancel in pairs[5] and we find that the divergence of the curl of *any vector function* $\mathbf{A}(\mathbf{r})$ vanishes identically,

$$\nabla \cdot (\nabla \times \mathbf{A}(\mathbf{r})) \equiv 0. \tag{3.10}$$

3.3   Components of the Gradient

Now let us make a geometrical interpretation of the gradient. The vector $\nabla \psi$ at an arbitrary point located at **r** relative to O is illustrated in Fig. 3.3. Consider a small displacement $d\mathbf{r}$ from **r**. The scalar product of $\nabla \psi$ and $d\mathbf{r}$ is seen to be

$$\nabla \psi \cdot d\mathbf{r} = (\nabla \psi)_x \, dx + (\nabla \psi)_y \, dy + (\nabla \psi)_z \, dz$$

$$= \frac{\partial \psi}{\partial x} \, dx + \frac{\partial \psi}{\partial y} \, dy + \frac{\partial \psi}{\partial z} \, dz$$

$$= |\nabla \psi||d\mathbf{r}| \cos \alpha$$

$$= ds |\nabla \psi| \cos \alpha,$$

where $ds \equiv |d\mathbf{r}|$.

---

[4]A catalogue of vector identities is given in Appendix A.

[5]For the functions we consider, the results are independent of the order in which differentiations are performed.

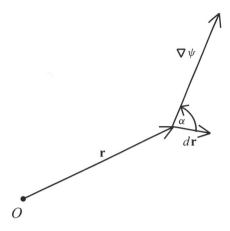

FIGURE 3.3. Geometrical interpretation of the gradient.

According to the rules for the differential of a function of several variables (See Eq. (3.1).), we also have

$$d\psi = \frac{\partial \psi}{\partial x} dx + \frac{\partial \psi}{\partial y} dy + \frac{\partial \psi}{\partial z} dz.$$

So,

$$d\psi = \nabla \psi \cdot d\mathbf{r}. \tag{3.11}$$

We also see that

$$d\psi = ds |\nabla \psi| \cos \alpha,$$

or

$$\frac{d\psi}{ds} = |\nabla \psi| \cos \alpha,$$

which is clearly the projection of the vector $\nabla \psi$ onto $d\mathbf{r}$. This quantity $d\psi/ds$ is called the *directional derivative*. Specifically, the directional derivative of a scalar function is defined to be the space rate of change of the function in a particular direction. Here $d\mathbf{r}$ is an infinitesimal displacement in the direction considered.

We see that the directional derivative $d\psi/ds$ has its maximum value when $\alpha = 0$, i.e., when the displacement $d\mathbf{r}$ is parallel to $\nabla \psi$. Thus, $\nabla \psi$ is a vector whose direction is that in which ψ has its greatest rate of change, and whose magnitude is equal to the directional derivative in this direction.

These results allow us to identify the components of $\nabla \psi$ in *any* coordinate system, if $d\mathbf{r}$ is known in the system. We consider two examples.

3.3.1 RECTANGULAR COORDINATES

As a check on this method, we use it to reconstruct the components of $\nabla\psi$ in rectangular coordinates. From Eq. (3.11), we have

$$d\psi = \frac{\partial\psi}{\partial x}\,dx + \frac{\partial\psi}{\partial y}\,dy + \frac{\partial\psi}{\partial z}\,dz$$
$$= \nabla\psi \cdot d\mathbf{r}$$
$$= (\nabla\psi)_x\,dx + (\nabla\psi)_y\,dy + (\nabla\psi)_z\,dz.$$

Since the differentials are independent, we see that

$$(\nabla\psi)_x = \frac{\partial\psi}{\partial x} \qquad (\nabla\psi)_y = \frac{\partial\psi}{\partial y} \qquad (\nabla\psi)_z = \frac{\partial\psi}{\partial z},$$

as we saw earlier from the direct application of the operator ∇ in Eq. (3.2) to the function $\psi(\mathbf{r})$.

3.3.2 SPHERICAL POLAR COORDINATES

We have used the vector $\mathbf{r}$ to represent the point P with rectangular coordinates (x, y, z). We can also specify the position of P by giving its distance and direction relative to the origin of the coordinate system. The distance we represent by $r = |\mathbf{r}|$ and the direction by two angles θ and ϕ. These parameters, called *spherical polar coordinates*, are defined in Fig. 3.4 in which their relation to the rectangular coordinates can be seen to be

$$x = r \sin\theta \cos\phi,$$
$$y = r \sin\theta \sin\phi,$$
$$z = r \cos\theta.$$

At the point P, we define three mutually perpendicular unit vectors $(\hat{\mathbf{r}}, \hat{\boldsymbol{\theta}}, \hat{\boldsymbol{\phi}})$, where the hat ($\hat{}$) indicates a unit vector:

- $\hat{\mathbf{r}}$ is in the direction of the position vector $\mathbf{r}$
- $\hat{\boldsymbol{\theta}}$ is in the direction in which the polar angle θ increases at P
- $\hat{\boldsymbol{\phi}}$ is in the direction in which the angle ϕ increases at P

Note that in a given coordinate system, the directions of the unit cartesian vectors $(\mathbf{i}, \mathbf{j}, \mathbf{k})$ do not change in moving from one point to another. By contrast, the directions of two or more of the polar vectors $(\hat{\mathbf{r}}, \hat{\boldsymbol{\theta}}, \hat{\boldsymbol{\phi}})$ may be different for two different points in space.

We can identify the components of the infinitesimal displacement $d\mathbf{r}$ in spherical polar coordinates by considering three successive displacements as indicated in Fig. 3.5:

1. $r \sin\theta\, d\phi$ (in the direction $\hat{\boldsymbol{\phi}}$)
2. $r\, d\theta$ (in the direction $\hat{\boldsymbol{\theta}}$)
3. dr (in the direction $\hat{\mathbf{r}}$)

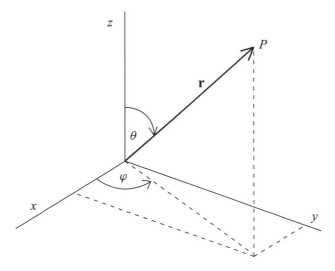

FIGURE 3.4. Relation between rectangular and spherical polar coordinates.

Thus, for the total infinitesimal displacement, we have

$$d\mathbf{r} = \hat{\mathbf{r}}\, dr + \hat{\boldsymbol{\theta}}\, r\, d\theta + \hat{\boldsymbol{\phi}}\, r\sin\theta\, d\phi. \tag{3.12}$$

Applying Eq. (3.11) again, we get

$$d\psi = \frac{\partial \psi}{\partial r}\, dr + \frac{\partial \psi}{\partial \theta}\, d\theta + \frac{\partial \psi}{\partial \phi}\, d\phi$$

$$= \nabla\psi \cdot d\mathbf{r}$$

$$= (\nabla\psi)_r dr + (\nabla\psi)_\theta r\, d\theta + (\nabla\psi)_\phi r \sin\theta\, d\phi,$$

from which we make the identifications

$$(\nabla\psi)_r = \frac{\partial\psi}{\partial r} \qquad (\nabla\psi)_\theta = \frac{1}{r}\frac{\partial\psi}{\partial\theta} \qquad (\nabla\psi)_\phi = \frac{1}{r\sin\theta}\frac{\partial\psi}{\partial\phi}.$$

3.3.3   AN APPLICATION IN PHYSICS

A particle of mass m lies outside a large, spherically symmetric body of mass M. Relative to a point at infinity, the gravitational potential energy V of the particle at a distance r from the center of the large mass is equal to the work done in moving it from the point at infinity to the point at the distance r from the center of the large body. The gravitational force on m is given by Newton's law of universal gravitation

$$\mathbf{F} = -\frac{GMm}{r^3}\mathbf{r}, \tag{3.13}$$

where $\mathbf{r}$ gives the position of m relative to the center of the large body as illustrated

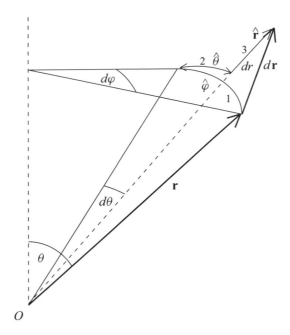

FIGURE 3.5. The differential $d\mathbf{r}$ in spherical polar coordinates.

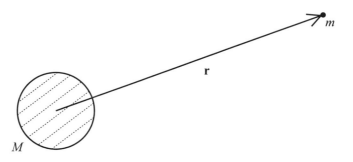

FIGURE 3.6. A particle in the gravitational field of a large body.

in Fig. 3.6. The minus sign in Eq. (3.13) reflects the fact that the gravitational force is attractive, i.e., the force on m is in the direction opposite to $\mathbf{r}$. Thus, for the gravitational potential energy of m we have

$$V(r) = -\int_\infty^r \mathbf{F}(\mathbf{r}') \cdot d\mathbf{r}' = -\int_\infty^r -\frac{GMm}{r'^3}\mathbf{r}' \cdot d\mathbf{r}'. \qquad (3.14)$$

From Eq. (3.12) and the definition of the unit vector $\hat{\mathbf{r}} = \mathbf{r}/r$ it follows that

$$\mathbf{r}' \cdot d\mathbf{r}' = r'\hat{\mathbf{r}}' \cdot (\hat{\mathbf{r}}' dr' + \hat{\boldsymbol{\theta}}' r' d\theta' + \hat{\boldsymbol{\phi}}' r' \sin\theta' d\phi') = r' dr'.$$

Substituting this result into Eq. (3.14), we get

$$V(r) = \int_\infty^r \frac{GMm}{r'^3} r' dr' = -\frac{GMm}{r'}\bigg|_\infty^r = -\frac{GMm}{r}. \tag{3.15}$$

Invoking Eq. (3.4), we see that Eqs. (3.13) and (3.15) are related by

$$\mathbf{F} = -\frac{GMm}{r^3}\mathbf{r} = \nabla\left(\frac{GMm}{r}\right) = -\nabla V(r).$$

That is, the gravitational force on the particle of mass m can be expressed in terms of the gradient of the corresponding gravitational potential energy,

$$\mathbf{F} = -\nabla V(r).$$

The gravitational force is an example of a *conservative* force. In general, any force $\mathbf{F}$ that is derivable from a potential energy $V(\mathbf{r})$ according to

$$\mathbf{F} = -\nabla V(\mathbf{r}) \tag{3.16}$$

is said to be a conservative force. It is clear from Eq. (3.9) that for a conservative force $\mathbf{F}$,

$$\nabla \times \mathbf{F} = 0. \qquad \text{(conservative force)} \tag{3.17}$$

3.3.4   WORK DONE BY A CONSERVATIVE FORCE

Generally, the work done in moving a particle from one point to another in a force field depends on the path the particle takes between the two points. For example, the work done in moving a particle from point 1 to point 2 in a viscous fluid is different for different paths connecting the two points. This is illustrated in Fig. 3.7 where the two paths are labeled a and b. However, for a conservative force $\mathbf{F}$ given by Eq. (3.16), we have

$$W = -\int_1^2 \mathbf{F} \cdot d\mathbf{r} = \int_1^2 \nabla V(\mathbf{r}) \cdot d\mathbf{r}$$
$$= \int_1^2 \left(\frac{\partial V}{\partial x} dx + \frac{\partial V}{\partial y} dy + \frac{\partial V}{\partial z} dz\right) = \int_1^2 dV = V(\mathbf{r}_2) - V(\mathbf{r}_1).$$

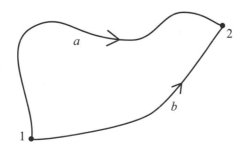

FIGURE 3.7. Work done in moving a particle along different paths.

Clearly, the work W depends only on the values of the potential energy at the end points $\mathbf{r}_1$ and $\mathbf{r}_2$ independent of the path traversed from $\mathbf{r}_1$ to $\mathbf{r}_2$. Hence, the work done moving a particle from one point to another in a conservative force field is independent of the path.

3.4   Flux

The *flux* of a vector quantity through a surface is a very useful concept in physics. A familiar example is the flux associated with a magnetic field. A magnetic field exists in a region of space in which a magnetic pole experiences a force. We make this notion quantitative by defining the field in terms of the force $\mathbf{F}$ that a small magnetic pole p experiences at a point in the magnetic field $\mathbf{B}$ according to

$$\mathbf{B} = \lim_{p \to 0} \frac{\mathbf{F}}{p}.$$

The direction of the magnetic field is in the direction of the force on a small N-pole. We could determine the pattern of the magnetic field in a given region by placing a small compass needle at various points in the space. For example, the field around a simple bar magnet could be traced out as illustrated in Fig. 3.8. It is clear that the compass needles, which indicate the direction of the field at the points where they are placed, align along directions tangent to the field lines. These lines that represent the magnetic field in a pictorial way are called lines of *magnetic flux*. They can also be used to represent the magnetic field quantitatively.

Consider the two closed loops labeled a and b in Fig. 3.8. The loops enclose equal areas, but loop b, located close to the magnet where the field is stronger, has more flux lines passing through it than does loop a, which is farther from the magnet where the field is weaker. Thus, we can take the *density* of magnetic flux to be representative of the *magnitude* of the magnetic field $\mathbf{B}$.

We have already identified the *direction* of $\mathbf{B}$ with the direction of the line tangent to the flux line at the field point.

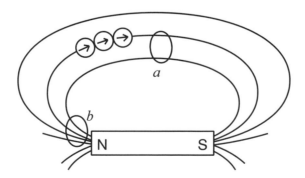

FIGURE 3.8. Magnetic field lines around a bar magnet.

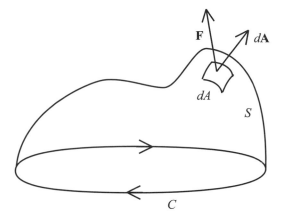

FIGURE 3.9. An open surface S bounded by a closed curve C.

The magnetic flux $d\Phi$ through an infinitesimal area element dA is defined by

$$d\Phi = B_\perp dA,$$

where $B_\perp$ is the component of the magnetic field $\mathbf{B}$ that is perpendicular to the area element dA. According to this definition, it is reasonable to interpret the magnetic field vector $\mathbf{B}$ as the *magnetic flux density* (flux per unit area). In fact, that designation is often used for $\mathbf{B}$ in the physics literature.

Now, let us generalize to an arbitrary vector field $\mathbf{F}$ and consider the corresponding flux passing through an open surface S that is bounded by a closed curve C as illustrated in Fig. 3.9. If we define $d\mathbf{A}$ to be a vector of magnitude dA with direction along the *outward* normal to the surface at dA, then we can write $d\Phi$ as the scalar product

$$d\Phi = \mathbf{F} \cdot d\mathbf{A}.$$

The *outward normal* to an open surface is defined as the direction in which a right-handed screw will advance if turned in the direction indicated by the arrows on the curve C which is the boundary of the surface.

The total flux of a vector field $\mathbf{F}$ through the surface S is

$$\Phi = \int_S \mathbf{F} \cdot d\mathbf{A}.$$

Consider an example from fluid mechanics in which a fluid of density $\rho(\mathbf{r}, t)$ moves with a flow velocity $\mathbf{v}(\mathbf{r}, t)$ as depicted in Fig. 3.10. The infinitesimal mass dm of fluid crossing dA in the time interval dt is equal to the mass of fluid in the cylinder of length $v\,dt$ shown in Fig. 3.10,

$$dm = \rho(v\,dt)(dA \cos\theta)$$
$$= \rho\,dt\,\mathbf{v} \cdot d\mathbf{A}.$$

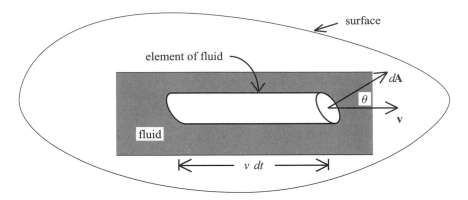

FIGURE 3.10. Fluid flowing through a closed surface.

Thus, the infinitesimal fluid current dI (rate of flow of fluid) is

$$dI = \frac{dm}{dt} = \rho \mathbf{v} \cdot d\mathbf{A}.$$

Integrating over the closed surface, we get

$$\int_{\text{surface}} dI = \int_{\text{surface}} \rho \mathbf{v} \cdot d\mathbf{A}.$$

The left-hand side of this equation represents the rate at which mass flows out through the surface (leaves the volume bounded by the surface). The right-hand side is the flux of the vector field $\rho \mathbf{v}$ through the closed surface.

3.4.1   THE DIVERGENCE THEOREM

Now, let us look at another way to calculate the net flux out over a *closed* surface. Consider a small rectangular box of volume $d^3r = dx\,dy\,dz$ with one corner at the point (x, y, z) as shown in Fig. 3.11. The flux out of the *front* of the box at $(x + dx, y, z)$ is

$$d\Phi_1 = F_x(x + dx, y, z)\,dA_x$$
$$= F_x(x + dx, y, z)\,dy\,dz.$$

The flux out of the *back* of the box at (x, y, z) is

$$d\Phi_2 = -F_x(x, y, z)\,dA_x$$
$$= -F_x(x, y, z)\,dy\,dz.$$

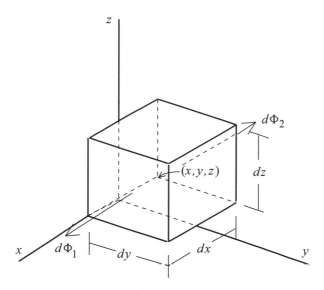

FIGURE 3.11. Infinitesimal rectangular volume element.

The *net* flux out of the box parallel to the x-axis is

$$d\Phi_1 + d\Phi_2 = [F_x(x + dx, y, z) - F_x(x, y, z)]\, dy\, dz$$
$$= \left(\frac{\partial F_x}{\partial x}\, dx\right) dy\, dz$$
$$= \frac{\partial F_x}{\partial x}\, d^3r.$$

With similar calculations for the other two pairs of faces of the box, we find that the *total outward flux* for the volume d^3r is

$$d\Phi = \left(\frac{\partial F_x}{\partial x} + \frac{\partial F_y}{\partial y} + \frac{\partial F_z}{\partial z}\right) d^3r$$
$$= \nabla \cdot \mathbf{F}\, d^3r.$$

Now, suppose we add up all infinitesimal contributions which make up a finite volume V. Consider two adjacent cells as shown in Fig. 3.12.

The *outward* flux at the back face of cell 1 is equal to the *inward* flux at the front face of cell 2. Therefore, upon calculating the *outward* flux from each cell, we find that these two contributions cancel. So, if we have a *finite* volume made up of many such cells, then the only place where there is no cancellation is on the outside faces of those cells at the surface of the volume V. The total outward flux for the finite volume V is then

$$\Phi - \int_V \nabla \cdot \mathbf{F}\, d^3r.$$

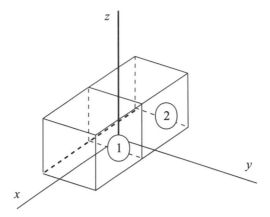

FIGURE 3.12. Contributions from adjacent elements.

But we have already seen that the outward flux through *any* closed surface S is, by definition

$$\Phi = \oint_S \mathbf{F} \cdot d\mathbf{A}.$$

From these last two results, we see that for any vector field $\mathbf{F}$ and any arbitrary volume V bounded by the closed surface S, we have

$$\oint_S \mathbf{F} \cdot d\mathbf{A} = \int_V \nabla \cdot \mathbf{F} \, d^3r. \tag{3.18}$$

This integral relation is known variously as the *divergence theorem* and as *Gauss's theorem*.

Example 1.

Let us verify that the divergence theorem holds for the vector field

$$\mathbf{F} = y^2 \mathbf{i} + xy \mathbf{j} + yz \mathbf{k}$$

and a volume which is a cube of side b as shown in Fig. 3.13.

Solution.

First, we calculate the left-hand side of Eq. (3.18). We do this surface integral piecewise for the six faces of the cube separately.

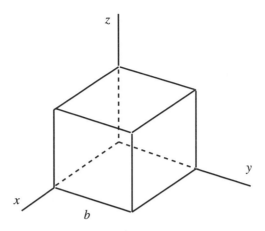

FIGURE 3.13. A cubical volume.

top : $\displaystyle\int_{z=b} (\mathbf{k}\, yz) \cdot (\mathbf{k}\, dx\, dy) = \int_0^b \int_0^b [yb]\, dx\, dy = \frac{1}{2}b^4.$

bottom : $\displaystyle\int_{z=0} (\mathbf{k}\, yz) \cdot (-\mathbf{k}\, dx\, dy) = 0.$

front : $\displaystyle\int_{x=b} (\mathbf{i}\, y^2) \cdot (\mathbf{i}\, dy\, dz) = \frac{1}{3}b^4.$

back : $\displaystyle\int_{x=0} (\mathbf{i}\, y^2) \cdot (-\mathbf{i}\, dy\, dz) = -\frac{1}{3}b^4.$

right : $\displaystyle\int_{y=b} (\mathbf{j}\, xy) \cdot (\mathbf{j}\, dx\, dz) = \frac{1}{2}b^4.$

left : $\displaystyle\int_{y=0} (\mathbf{j}\, xy) \cdot (-\mathbf{j}\, dx\, dz) = 0.$

Adding up these contributions, we get

$$\oint_S \mathbf{F} \cdot d\mathbf{A} = b^4.$$

For the right-hand side of Eq. (3.18), we have

$$\int_V \nabla \cdot \mathbf{F}\, d^3r = \int_0^b \int_0^b \int_0^b (x+y)\, dx\, dy\, dz = b^4.$$

Clearly, Eq. (3.18) is verified for this vector field and the volume in Fig. 3.13.

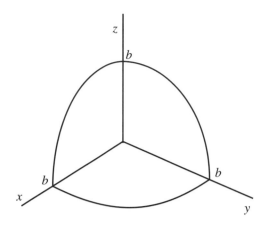

FIGURE 3.14. An octant of a sphere.

Example 2.

In spherical polar coordinates, a vector field $\mathbf{F}$ is given by

$$\mathbf{F} = \hat{\mathbf{r}}(r \cos \theta \cos \phi) + \hat{\boldsymbol{\theta}} (r \cos \theta \sin \phi) + \hat{\boldsymbol{\phi}} (r \sin \theta \cos \phi).$$

Check the validity of the divergence theorem for this vector and the volume that is one octant of a sphere of radius b, as illustrated in Fig. 3.14.

Solution.

The surface which forms the boundary of the volume can be divided into four parts. The corresponding area elements are:

$$\text{spherical surface:} \quad d\mathbf{A} = \hat{\mathbf{r}} b^2 \sin \theta \, d\theta d\phi \quad (r = b)$$
$$xz\text{-plane:} \quad d\mathbf{A} = -\hat{\boldsymbol{\phi}} r \, dr \, d\theta \quad (\phi = 0)$$
$$xy\text{-plane:} \quad d\mathbf{A} = \hat{\boldsymbol{\theta}} r \, dr \, d\phi \quad \left(\theta = \tfrac{\pi}{2}\right)$$
$$yz\text{-plane:} \quad d\mathbf{A} = \hat{\boldsymbol{\phi}} r \, dr \, d\theta \quad \left(\phi = \tfrac{\pi}{2}\right)$$

On integrating over the four pieces of the surface, we find that

$$\oint_S \mathbf{F} \cdot d\mathbf{A} = \tfrac{1}{6} b^3.$$

In spherical polar coordinates, the divergence of an arbitrary vector function $\mathbf{A}(\mathbf{r})$ is given by[6]

$$\nabla \cdot \mathbf{A} = \frac{1}{r^2} \frac{\partial}{\partial r} (r^2 A_r) + \frac{1}{r \sin \theta} \frac{\partial}{\partial \theta} (\sin \theta A_\theta) + \frac{1}{r \sin \theta} \frac{\partial A_\phi}{\partial \phi}.$$

So, for the divergence of $\mathbf{F}$, we get

$$\nabla \cdot \mathbf{F} = 3 \cos \theta \cos \phi + \frac{(1 - 2 \sin^2 \theta) \sin \phi}{\sin \theta} - \sin \phi.$$

---

[6]See Appendix A.3.2.

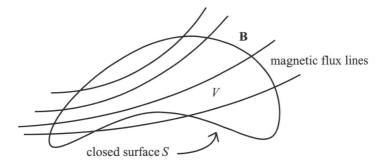

FIGURE 3.15. A volume V bounded by a closed surface S.

With the spherical volume element $d^3r = r^2 \sin\theta\, dr\, d\theta\, d\phi$, this leads to

$$\int_V \nabla \cdot \mathbf{F}\, d^3r = \int_0^{\pi/2} \int_0^{\pi/2} \int_0^b \nabla \cdot \mathbf{F}\, r^2 dr \sin\theta\, d\theta\, d\phi = \tfrac{1}{6} b^3$$

and the validity of the divergence theorem is verified.

3.4.2 THE MAGNETIC FIELD AGAIN

The Biot-Savart law is an empirical result which was obtained in the nineteenth century from the direct measurement of magnetic forces that electric currents exert on each other. Two of Maxwell's equations follow from this law. One of these is[7]

$$\nabla \cdot \mathbf{B} = 0, \tag{3.19}$$

where $\mathbf{B}$ is the magnetic flux density.

An arbitrary closed surface S is constructed in a magnetic field $\mathbf{B}$ as shown in Fig. 3.15. The net magnetic flux Φ_{mag} passing outward through the surface S is given by

$$\Phi_{\text{mag}} = \oint_S \mathbf{B} \cdot d\mathbf{A} = \int_V \nabla \cdot \mathbf{B}\, d^3r = 0,$$

where we have invoked the divergence theorem and the Maxwell equation $\nabla \cdot \mathbf{B} = 0$. Thus,

$$\Phi_{\text{mag}} = \oint_S \mathbf{B} \cdot d\mathbf{A} = 0 \qquad \text{always!}$$

That is, the net magnetic flux out through any closed surface is always zero. Magnetic flux lines are closed loops. The implication is that there exist no isolated sources of magnetic flux. According to Maxwell's equations, there exist no magnetic charges (monopoles) in the universe. Magnetic poles come in pairs (dipoles): an N-pole accompanied by an S-pole.

---

[7]See Appendix B.

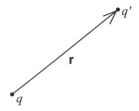

FIGURE 3.16. The Coulomb force one electric charge exerts on another.

3.4.3 AN APPLICATION OF THE DIVERGENCE THEOREM

Two electric charges q and q' are separated by a distance r as shown in Fig. 3.16. According to Coulomb's law, the force $\mathbf{F}$ that q exerts on q' is

$$\mathbf{F} = k\frac{qq'}{r^2}\hat{\mathbf{r}}.$$

The electric field $\mathbf{E}$ at the position of q' due to q is given by

$$\mathbf{E} = \lim_{q' \to 0}\frac{\mathbf{F}}{q'} = \lim_{q' \to 0}\frac{1}{q'}\left(k\frac{qq'}{r^2}\hat{\mathbf{r}}\right) = \frac{kq}{r^2}\hat{\mathbf{r}}.$$

Now calculate the surface integral of the normal component of $\mathbf{E}$ over a closed surface S enclosing the charge q.

From Fig. 3.17, we see that the area element $d\mathbf{A}$ on this surface S is

$$d\mathbf{A} = \mathbf{n}\,dA,$$

where $\mathbf{n}$ is a unit vector along the *outward* normal to the surface S. Thus,

$$\oint_S \mathbf{E} \cdot d\mathbf{A} = \oint_S \left(\frac{kq}{r^2}\hat{\mathbf{r}}\right) \cdot \mathbf{n}\,dA$$

$$= kq\oint \frac{\hat{\mathbf{r}} \cdot \mathbf{n}}{r^2}\,dA.$$

It is also clear from Fig. 3.17 that $\hat{\mathbf{r}} \cdot \mathbf{n} = \cos\alpha$, where α is the angle between the element dA of S and an element of a sphere of radius r. That is, the area element $\hat{\mathbf{r}} \cdot \mathbf{n}\,dA$ is equal to the projection of dA onto a sphere of radius r or $\cos\alpha\,dA = r^2 \sin\theta\,d\theta\,d\phi$. From this, it follows that

$$\oint_S \frac{\hat{\mathbf{r}} \cdot \mathbf{n}}{r^2}\,dA = \oint_S \frac{\cos\alpha\,dA}{r^2} = \int_0^{2\pi}\int_0^{\pi}\sin\theta\,d\theta d\phi \equiv \int d\Omega = 4\pi,$$

where $d\Omega$ denotes the infinitesimal solid angle given by $d\Omega = \sin\theta\,d\theta\,d\phi$ in spherical polar coordinates. Therefore, for the surface integral, we obtain

$$\oint_S \mathbf{E} \cdot d\mathbf{A} = 4\pi kq.$$

Now suppose q lies outside the closed surface S. In this case, each infinitesimal

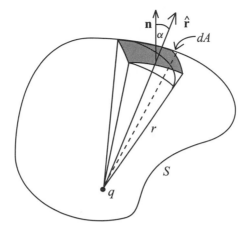

FIGURE 3.17. An arbitrary surface S enclosing an electric charge q.

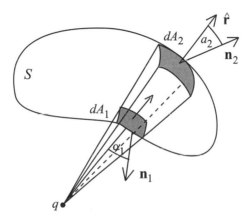

FIGURE 3.18. An electric charge q located outside a closed surface S.

solid angle $d\Omega$ intercepts S twice as illustrated in Fig. 3.18. Thus, the infinitesimal elements contribute to the surface integral in pairs. For example, the pair shown in Fig. 3.18 yields,

$$\mathbf{E}_1 \cdot d\mathbf{A}_1 + \mathbf{E}_2 \cdot d\mathbf{A}_2 = \left(\frac{kq}{r_1^2}\hat{\mathbf{r}}_1\right) \cdot \mathbf{n}_1 \, dA_1 + \left(\frac{kq}{r_2^2}\hat{\mathbf{r}}_2\right) \cdot \mathbf{n}_2 \, dA_2.$$

In Fig. 3.18, we also see that

$$\hat{\mathbf{r}}_1 \cdot \mathbf{n}_1 = \cos(\pi - \alpha_1) = -\cos\alpha_1.$$

So, the two contributions may be expressed as

$$\left(-\frac{kq\cos\alpha_1}{r_1^2}dA_1\right) + \left(\frac{kq\cos\alpha_2}{r_2^2}dA_2\right)$$
$$= kq(-d\Omega_1 + d\Omega_2) = 0,$$

where the last equality follows from the fact that the same solid angle intercepts both pieces of the surface $(d\Omega_1 = d\Omega_2)$. This leads to the result,

$$\oint_S \mathbf{E} \cdot d\mathbf{A} = \begin{cases} 4\pi kq & \text{for } q \text{ inside } S \\ 0 & \text{for } q \text{ outside } S \end{cases}$$

For a system of N charges q_i, we have

$$\oint \mathbf{E} \cdot d\mathbf{A} = \oint \left(\sum_{i=1}^{N} \mathbf{E}_i \right) \cdot d\mathbf{A}$$

$$= 4\pi k \sum_{i} q_i = 4\pi k \times (\text{charge inside } S).$$
$$\text{\footnotesize (for } q_i \text{ inside } S)$$

A continuous distribution of electric charge is represented by a charge density $\rho(\mathbf{r})$. The total charge Q in such a distribution is then given by

$$Q = \int_V \rho(\mathbf{r}) \, d^3r.$$

Invoking the divergence theorem, we have

$$\oint_S \mathbf{E} \cdot d\mathbf{A} = \int_V \nabla \cdot \mathbf{E} \, d^3r = 4\pi k Q_{\text{enclosed}} = 4\pi k \int_V \rho(\mathbf{r}) \, d^3r.$$

On rearranging,

$$\int_V \left(\nabla \cdot \mathbf{E} - 4\pi k \rho(\mathbf{r}) \right) d^3r = 0.$$

Finally, since S (hence V) is arbitrary, we have the general result,

$$\nabla \cdot \mathbf{E} = 4\pi k \rho(\mathbf{r}). \tag{3.20}$$

Let us compare this result with the corresponding relation for a magnetic field Eq. (3.19). By examining the deflection of a small positive point charge in an electric field $\mathbf{E}$ analogous to the way we used a small compass to investigate the magnetic field around a bar magnet, we can map out the distribution of electric flux lines in this electric field. Electric flux lines extend radially out (in the direction of the electric field) from a positive point charge as shown in Fig. 3.19. Upon surrounding this point charge with an arbitrary closed surface S, we see from Fig. 3.19 that there is a net electric flux out over the surface S. Therefore, the right-hand side of Eq. (3.20) cannot be zero in this case. In general, the right-hand side of this equation is proportional to the electric charge density. If magnetic charges existed in our universe, then Eq. (3.19) would have to be modified so that the right-hand side would be proportional to the *magnetic* charge density.

3.4.4   STOKES'S THEOREM

In this section, we obtain another useful integral theorem that has wide application in theoretical physics. In preparation for proving this theorem, we introduce the

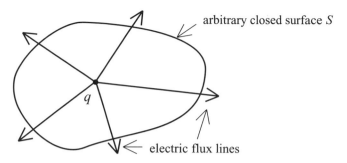

FIGURE 3.19. Electric flux lines due to an electric charge q.

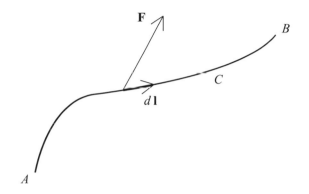

FIGURE 3.20. Integrating along a line.

notion of a *line integral*. The line integral of a vector $\mathbf{F}$ over a curve C is the product of the line element dl and the component of $\mathbf{F}$ along $d\mathbf{l}$ integrated over the line from point A to point B, as illustrated in Fig. 3.20. That is,

$$\int_C \mathbf{F} \cdot d\mathbf{l}.$$

If the end points A and B coincide so that the curve C forms a closed path, then we write the line integral as

$$\oint_C \mathbf{F} \cdot d\mathbf{l}.$$

Let us consider an infinitesimally small rectangular loop parallel to the yz-plane with one corner of the loop at the point (x, y, z) as shown in Fig. 3.21. We calculate the line integral of $\mathbf{F}$ around this small loop as indicated in the drawing. Calculating the four pieces in the order shown, we obtain

$$\oint_C \mathbf{F} \cdot d\mathbf{l} = F_y(x, y, z)\,dy + F_z(x, y + dy, z)\,dz$$
$$+ F_y(x, y, z + dz)(-dy) + F_z(x, y, z)(-dz).$$

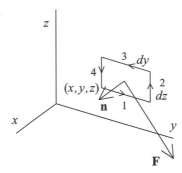

FIGURE 3.21. Infinitesimal area element bounded by a closed loop.

Collecting the terms, we have

$$\oint_C \mathbf{F} \cdot d\mathbf{l} = [F_z(x, y + dy, z) - F_z(x, y, z)]dz$$
$$- [F_y(x, y, z + dz) - F_y(x, y, z)]dy.$$

Recognizing the quantities in the square brackets as partial derivatives, we write

$$\oint_C \mathbf{F} \cdot d\mathbf{l} = \left[\frac{\partial F_z}{\partial y} dy\right] dz - \left[\frac{\partial F_y}{\partial z} dz\right] dy$$
$$= (\nabla \times \mathbf{F})_x \, dA_x = (\nabla \times \mathbf{F}) \cdot \mathbf{n} \, dA$$
$$= (\nabla \times \mathbf{F}) \cdot d\mathbf{A}. \qquad (3.21)$$

If we consider a finite surface S to consist of a large number of such small loops as illustrated in Fig. 3.22 a, we see that contributions to the line integral from adjoining parts of adjacent loops cancel. For example, in Fig. 3.22 b we have four of these loops. Note that along the line between loops 1 and 2 the contributions to the line integrals have the same magnitude but the opposite sense, so when the line integrals are added together these contributions will cancel. It is clear that similar cancellations will occur along the boundaries of all contiguous loops. The only places where no cancellations occur are along those parts of loops which coincide

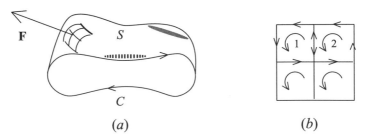

FIGURE 3.22. Contributions from infinitesimal area elements.

with the bounding curve C in Fig. 3.22 a. Thus,

$$\sum_{\text{all loops}} \oint \mathbf{F} \cdot d\mathbf{l} = \oint_C \mathbf{F} \cdot d\mathbf{l}.$$

Now, we also see from Eq. (3.21) that

$$\sum_{\text{all loops}} \oint \mathbf{F} \cdot d\mathbf{l} = \sum_{\text{all loops}} (\nabla \times \mathbf{F}) \cdot d\mathbf{A}$$

$$= \int_S (\nabla \times \mathbf{F}) \cdot d\mathbf{A}.$$

Finally, we obtain the result

$$\oint_C \mathbf{F} \cdot d\mathbf{l} = \int_S (\nabla \times \mathbf{F}) \cdot d\mathbf{A}. \tag{3.22}$$

This integral relation is known as *Stokes's theorem. The line integral of a vector* **F** *around any closed curve C is equal to the surface integral of the normal component of the curl of* **F** *over any open surface S for which C is a boundary.*

In verifying Stokes's theorem for closed curves that lie in a plane, it will be useful to be able to construct a unit vector perpendicular to such a plane.[8] We have seen that if the vectors **a** and **b** are not colinear, then $\mathbf{a} \times \mathbf{b}$ is a vector perpendicular to the plane defined by **a** and **b**. We can use this property of the cross product to construct a unit vector **n** perpendicular to (or *normal* to) any arbitrary plane. For example, let us construct the unit vector **n** normal to the plane shown in Fig. 3.23.

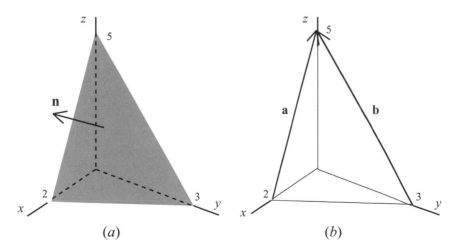

(a) $\qquad\qquad\qquad\qquad\qquad$ (b)

FIGURE 3.23. A unit vector perpendicular to a plane.

[8]See D. J. Griffiths, *Introduction to Electrodynamics*, Prentice-Hall, Englewood Cliffs, NJ, 1981, p. 12.

With the vectors **a** and **b** defined as shown in Fig. 3.23 *b*, it is clear that

$$\mathbf{n} = \frac{\mathbf{a} \times \mathbf{b}}{|\mathbf{a} \times \mathbf{b}|}.$$

From Fig. 3.23 *b*, we see that

$$\mathbf{a} = -2\,\mathbf{i} + 0\,\mathbf{j} + 5\,\mathbf{k}$$
$$\mathbf{b} = 0\,\mathbf{i} - 3\,\mathbf{j} + 5\,\mathbf{k}.$$

From these results, we get

$$\mathbf{a} \times \mathbf{b} = 15\,\mathbf{i} + 10\,\mathbf{j} + 6\,\mathbf{k}$$

and

$$|\mathbf{a} \times \mathbf{b}| = 19.$$

Thus,

$$\mathbf{n} = \frac{15}{19}\,\mathbf{i} + \frac{10}{19}\,\mathbf{j} + \frac{6}{19}\,\mathbf{k}.$$

Example 3.
Verify Stokes's theorem for the vector field

$$\mathbf{F} = y\,\mathbf{i} + z\,\mathbf{j}$$

and the surface represented by the shaded triangular area in Fig. 3.24.

Solution.
Writing the curl of **F** in rectangular coordinates leads to

$$\nabla \times \mathbf{F} = -\mathbf{i} - \mathbf{k}.$$

From the cross product, we get the unit vector **n** perpendicular to the surface

$$\mathbf{n} = \frac{6}{7}\,\mathbf{i} + \frac{2}{7}\,\mathbf{j} + \frac{3}{7}\,\mathbf{k}.$$

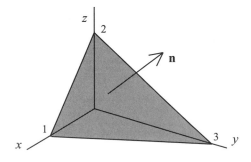

FIGURE 3.24. Surface for Example 3.

Using these results in Eq. (3.22), the surface integral becomes

$$\int_S (\nabla \times \mathbf{F}) \cdot \mathbf{n} \, dA = -\frac{9}{7} A.$$

From simple geometry, the area A of the shaded triangle is found to be $7/2$. Finally, we find that the surface integral has the value

$$\int_S (\nabla \times \mathbf{F}) \cdot \mathbf{n} \, dA = -\frac{9}{2}.$$

The line integral of $\mathbf{F}$ around the closed loop which is the boundary of the shaded triangle in Fig. 3.24 consists of three pieces,

$$\oint \mathbf{F} \cdot d\mathbf{l} = \int_{(z=0)} (y \, dx + z \, dy) + \int_{(x=0)} z \, dy + \int_{(y=0)} y \, dx$$
$$= \int_1^0 (-3x + 3) \, dx + \int_3^0 (-\tfrac{2}{3}y + 2) \, dy + 0 = -\frac{9}{2},$$

which is the same value we obtained for the surface integral. Thus, Stokes's theorem is verified.

Example 4.

As an application in spherical polar coordinates, check Stokes's theorem for the vector field,

$$\mathbf{F} = \hat{\mathbf{r}} \, (r \cos \theta \cos^2 \phi) + \hat{\boldsymbol{\theta}} \, (r \cos^2 \theta) + \hat{\boldsymbol{\phi}} \, (r \sin^2 \theta)$$

using the spherical cap in Fig. 3.25.

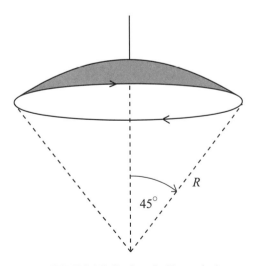

FIGURE 3.25. Surface for Example 4.

Solution.

Integrating around the circular boundary to the cap as illustrated in Fig. 3.25, we get

$$\oint \mathbf{F} \cdot d\mathbf{l} = \oint \mathbf{F} \cdot \hat{\phi} R \sin\frac{\pi}{4} d\phi = \frac{\pi R^2}{\sqrt{2}}.$$

To calculate the surface integral, we use the area element

$$d\mathbf{A} = \hat{\mathbf{r}} R^2 \sin\theta \, d\theta d\phi.$$

In spherical polar coordinates, the curl of any vector $\mathbf{F}$ is given by[9]

$$\nabla \times \mathbf{F} = \hat{\mathbf{r}}\frac{1}{r\sin\theta}\left[\frac{\partial}{\partial\theta}(\sin\theta \, F_\phi) - \frac{\partial F_\theta}{\partial\phi}\right]$$

$$+ \hat{\theta}\frac{1}{r}\left[\frac{1}{\sin\theta}\frac{\partial F_r}{\partial\phi} - \frac{\partial}{\partial r}(r F_\phi)\right] + \hat{\phi}\frac{1}{r}\left[\frac{\partial}{\partial r}(r F_\theta) - \frac{\partial F_r}{\partial\theta}\right].$$

For the vector in this example, this expression yields

$$\nabla \times \mathbf{F} = \hat{\mathbf{r}}(3\sin\theta\cos\theta) + \hat{\theta}(-2\sin\phi\cos\phi - 2\sin^2\theta)$$

$$+ \hat{\phi}(2\cos^2\theta - \cos\theta\cos^2\phi).$$

With these results, we obtain for the surface integral

$$\int (\nabla \times \mathbf{F}) \cdot d\mathbf{A} = \int_0^{2\pi}\int_0^{\pi/4}\left[\hat{\mathbf{r}}(3\sin\theta\cos\theta)\right] \cdot (\hat{\mathbf{r}}R^2\sin\theta d\theta \, d\phi) = \frac{\pi R^2}{\sqrt{2}}.$$

Again, Stokes's theorem is verified.

3.4.5   AN APPLICATION OF STOKES'S THEOREM

Ampere's law is another empirical law from electrodynamics. We use it to illustrate an application of Stokes's theorem. We have seen that a magnetic field $\mathbf{B}$ is defined as a region in which a magnetic pole p experiences a force. The force $\mathbf{F}$ exerted on p by the field $\mathbf{B}$ is

$$\mathbf{F} = p\mathbf{B}.$$

Coulomb found that the force $\mathbf{F}$ that a magnetic pole p exerts on a second pole p' is[10]

$$\mathbf{F} = k'\frac{pp'}{r^2}\hat{\mathbf{r}}, \tag{3.23}$$

where $\mathbf{r}$ is the vector that gives the position of p' relative to p. Remember, as far as we know, magnetic monopoles do not exist. To carry out this measurement in

---

[9]See Appendix A.3.2.
[10]In mks units, the constant k' has the value $k' = 10^{-7}$ N·s²/C².

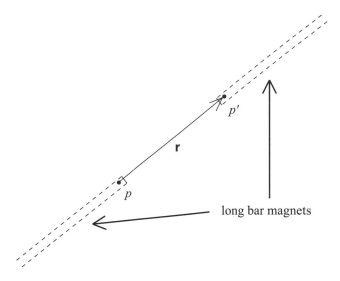

FIGURE 3.26. The force one magnetic pole exerts on another.

practice, one could use very long bar magnets as illustrated in Fig. 3.26. With this arrangement, the pole at the opposite end of the magnet with pole p will be far enough away to make only a small contribution to the force on p'.

Magnetic fields are produced by electric currents. For example, for points not too near the ends of a long, straight, current-carrying wire, the magnitude of the magnetic field at a perpendicular distance r from the wire, as shown in Fig. 3.27a, is given by

$$B = \frac{2k'I}{r}.$$

The direction of this magnetic field is tangent to a circle centered on the current I in the sense of a right-handed screw turning in the direction of the field and advancing in the direction of the current. This is seen by looking at the current end-on as it comes out of the plane of the page in Fig. 3.27 b, as opposed to Fig. 3.27 a

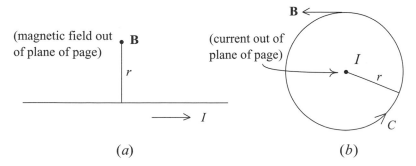

FIGURE 3.27. Magnetic field around a current-carrying wire.

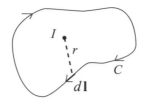

FIGURE 3.28. A closed loop surrounding a current.

where the current lies *in* the plane of the page. The work done in carrying a magnetic pole p around the circular path C in the magnetic field $\mathbf{B}$ due to the current I as shown in Fig. 3.27b is

$$W = -\oint_C \mathbf{F} \cdot d\mathbf{l} = -\oint_C p\mathbf{B} \cdot d\mathbf{l} = -p \int_0^{2\pi} \left(\frac{2k'I}{r}\right)(r \, d\phi) = -4\pi k' p I.$$

Dividing by p, we obtain the relation

$$\oint_C \mathbf{B} \cdot d\mathbf{l} = 4\pi k' I. \tag{3.24}$$

Note that the *size* of the circle is not specified. In fact (although we have not proved it here), the result in Eq. (3.24) is independent of the *shape* of the path C, as well. Therefore, for an arbitrary closed curve C surrounding the current I as shown in Fig. 3.28, the line integral of the magnetic field due to I integrated around C is given by Eq. (3.24). Applying these results to a distribution of N currents I_i, with the ith one making a contribution $\mathbf{B}_i$ to the field at $d\mathbf{l}$, we get

$$\oint_C \mathbf{B} \cdot d\mathbf{l} = \oint_C \sum_{i=1}^N \mathbf{B}_i \cdot d\mathbf{l} = \sum_{i=1}^N \oint_C \mathbf{B}_i \cdot d\mathbf{l} = \sum_{i=1}^N 4\pi k' I_i = 4\pi k' I_{\text{enclosed by } C}.$$

Thus, we arrive at Ampere's circuital law,

$$\oint_C \mathbf{B} \cdot d\mathbf{l} = 4\pi k' I_{\text{enclosed by } C},$$

where C is any arbitrary closed path.

Conservation of electric charge requires currents passing through C also to pass through an arbitrary surface S for which the closed path C is a boundary as illustrated in Fig. 3.29. It is convenient to define a current *density* $\mathbf{j}$ such that the total current through an arbitrary open surface S bounded by a closed curve C is given by

$$I_{\text{enclosed by } C} = \int_S \mathbf{j} \cdot d\mathbf{A}.$$

With this result, Eq. (3.24) may be written as

$$\oint_C \mathbf{B} \cdot d\mathbf{l} = 4\pi k' \int_S \mathbf{j} \cdot d\mathbf{A}.$$

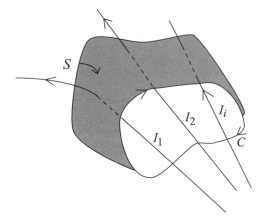

FIGURE 3.29. An arbitrary surface bounded by a closed loop.

Invoking Stokes's theorem and rearranging, we get

$$\int_S (\nabla \times \mathbf{B} - 4\pi k' \mathbf{j}) \cdot d\mathbf{A} = 0.$$

Since S is arbitrary, the integrand must vanish, leading to the final result

$$\nabla \times \mathbf{B} = 4\pi k' \mathbf{j}.$$

This is the differential form of Ampere's law for steady currents and fields in nonmagnetic media. It is a special case of another one of Maxwell's equations.[11]

3.5   Exercises

1. By successive applications of the definition of the curl in rectangular coordinates, show that

$$\nabla \times (\nabla \times \mathbf{F}) = -\nabla^2 \mathbf{F} + \nabla(\nabla \cdot \mathbf{F}),$$

   where $\mathbf{F}$ is any vector function of the space coordinates.
2. For each of the following vector functions, calculate $\nabla \cdot \mathbf{F}$ and $\nabla \times \mathbf{F}$.

   a. $\mathbf{F} = z\,\mathbf{i} + x\,\mathbf{j} + y\,\mathbf{k}$
   b. $\mathbf{F} = yz\,\mathbf{i} + xz\,\mathbf{j} + xy\,\mathbf{k}$
   c. $\mathbf{F} = x^3\,\mathbf{i} + z^3\,\mathbf{j} + 3yz^2\,\mathbf{k}$

3. Prove that there is no vector $\mathbf{F}$ such that

$$\nabla \times \mathbf{F} = \mathbf{r},$$

   where $\mathbf{r} = \mathbf{i}\,x + \mathbf{j}\,y + \mathbf{k}\,z$.

---

[11] See Appendix B.

4. The behavior of a particle of mass m moving under the influence of a force $\mathbf{F}$ is described by Newton's second law of motion,

$$\mathbf{F} = m\mathbf{a}.$$

We apply the term "central force" to any force field whose magnitude depends only on the particle's *distance* from the center of force and *not* on its *direction*. The general form for such a force is

$$\mathbf{F} = \hat{\mathbf{r}}F(r)$$

where $\hat{\mathbf{r}}$ is a unit vector in the direction of the position vector $\mathbf{r}$. For a particle moving with velocity $\mathbf{v}$ in a central force field, show that

$$\frac{d}{dt}(\mathbf{r} \times \mathbf{v}) = 0.$$

Do you recognize the conservation law associated with this result?

5. Calculate the gradient of each of the following scalar fields.

a. $\psi(\mathbf{r}) = xyz$
b. $\psi(\mathbf{r}) = r^{\lambda}$ ($\lambda = $ constant)
c. $\psi(\mathbf{r}) = x + y + z$
d. $\psi(\mathbf{r}) = xy + xz + yz$

6. A particle of mass m is moving in one dimension under the influence of a force F given by

$$F = -ax + bx^2,$$

where x is the position of the particle. The parameters a and b are real, positive constants. At x, the velocity of the particle is v. Calculate the total energy of the particle. Assume the potential energy is zero at equilibrium. Make a sketch showing the potential energy of this particle as a function of x.

7. Calculate the divergence and the curl of the vector field

$$\mathbf{F} = \mathbf{i}\,(6xz^2 + 9xy^2) + \mathbf{j}\,(9x^2y + 9yz^2) + \mathbf{k}\,(6x^2z + 6y^2z).$$

Wherever possible, express your results in terms of the vector $\mathbf{r}$ and its magnitude r. In a sentence, describe completely the orientation of the curl relative to the rectangular coordinate system.

8. Given the position vector $\mathbf{r} = \mathbf{i}\,x + \mathbf{j}\,y + \mathbf{k}\,z$, the arbitrary vector $\mathbf{A}(\mathbf{r})$, and the arbitrary *constant* vector $\mathbf{B}$, use rectangular coordinates to verify each of the following:

a. $(\mathbf{A} \cdot \nabla)\mathbf{r} = \mathbf{A}$
b. $(\mathbf{A} \times \nabla) \cdot \mathbf{r} = 0$
c. $\nabla \cdot (\mathbf{A} \times \mathbf{r}) = \mathbf{r} \cdot (\nabla \times \mathbf{A})$
d. $\nabla(\mathbf{r} \cdot \mathbf{B}) = \mathbf{B}$

9. Verify the divergence theorem for each of the following vector fields and the rectangular box depicted in Fig. 3.30.

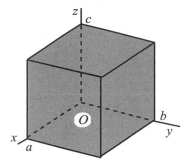

FIGURE 3.30. Surface for Exercise 3.9.

a. $\mathbf{F} = x(x + y)\mathbf{i} + x(y + z)\mathbf{j} + z^2\mathbf{k}$,
b. $\mathbf{F} = x^2 y\mathbf{i} + xy^2\mathbf{j} + xyz\mathbf{k}$
c. $\mathbf{F} = (2x + y)\mathbf{i} + (y + 3z)\mathbf{j} + (3x + 2z)\mathbf{k}$

10. Check the divergence theorem for each of the following vector fields using a hemispherical volume of radius R in the upper half plane with the center of curvature located at the origin of the coordinate system. The volume is closed at the bottom by a circular disk of radius R.

   a. $\mathbf{F} = \hat{\mathbf{r}}(r\cos\theta) + \hat{\boldsymbol{\theta}}(r\sin\theta) + \hat{\boldsymbol{\phi}}(r\sin\theta)$
   b. $\mathbf{F} = \hat{\mathbf{r}}(r^2\sin^2\theta) + \hat{\boldsymbol{\theta}}(r^2\sin\phi) + \hat{\boldsymbol{\phi}}(r^2\sin^2\phi)$

11. Consider a segment of a sphere of radius R with center of curvature at the origin of the coordinate system. This segment represents one fourth of the volume of the sphere and lies in the region where $y > 0$ and $z > 0$. Use this volume to check the divergence theorem for the following vector fields.

   a. $\mathbf{F} = \hat{\mathbf{r}}(r\cos\theta) + \hat{\boldsymbol{\theta}}(r\sin\theta) + \hat{\boldsymbol{\phi}}(r\sin\theta)$
   b. $\mathbf{F} = \hat{\mathbf{r}}(r^2\sin^2\theta) + \hat{\boldsymbol{\theta}}(r^2\sin\phi) + \hat{\boldsymbol{\phi}}(r^2\sin^2\phi)$
   c. $\mathbf{F} = \hat{\mathbf{r}}(r\sin^2\theta) + \hat{\boldsymbol{\theta}}(r\cos\phi) + \hat{\boldsymbol{\phi}}(r\sin\phi)$

12. Check the divergence theorem for the vector function

$$\mathbf{F} = \hat{\mathbf{r}}\, r\sin\theta\cos\theta + \hat{\boldsymbol{\theta}}\, r\sin\theta + \hat{\boldsymbol{\phi}}\, r\sin\theta\sin\phi$$

   using the segment of the sphere shown in Fig. 3.31.
   HINT: The area elements are

   a. top: $d\mathbf{A} = \hat{\mathbf{r}}\, dA$
   b. side: $d\mathbf{A} = \hat{\boldsymbol{\theta}}\, dA$.

   Find dA in each case. Show your work clearly. Make drawings as appropriate.

13. A sphere of radius R is cut into four equal parts. One piece is placed in a rectangular coordinate system such that one of the flat faces of the segment lies in the xz-plane and the other flat face is in the xy-plane. The center of curvature of the spherical surface is at the origin of this coordinate system. Looking down along the x-axis, the cross section of this piece appears as

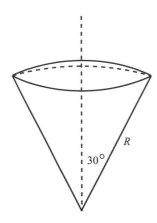

FIGURE 3.31. Surface for Exercise 3.12.

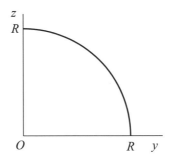

FIGURE 3.32. Surface for Exercise 3.13.

shown in Fig. 3.32. For this segment of the sphere and the vector

$$\mathbf{F} = \hat{\mathbf{r}}\,(r\cos^2\theta) + \hat{\boldsymbol{\theta}}\,(r\sin^2\theta) + \hat{\boldsymbol{\phi}}\,(r\sin\theta\cos\phi),$$

show that the divergence theorem holds.

14. Three noncolinear points are sufficient to define a plane. For each of the following sets of coordinates (x, y, z), make a sketch of the corresponding plane showing its relation to the rectangular coordinate system. For each plane, construct a *unit* vector $\mathbf{n}$ that is perpendicular to it.

   a. $(0, 0, 0)$ $(0, 15, 0)$ $(5, 0, 12)$
   b. $(4, 0, 0)$ $(0, 7, 0)$ $(0, 0, 3)$
   c. $(12, 16, 0)$ $(0, 0, 15)$ $(12, 0, 15)$
   d. $(13, 7, 9)$ $(13, 0, 0)$ $(0, 7, 9)$
   e. $(0, 0, 0)$ $(3, 5, 0)$ $(4, 1, 8)$

15. Verify Stokes's theorem for each of the following vector fields $\mathbf{F}$ using the plane surface shown in Fig 3.33.

   a. $\mathbf{F} = -3z\,\mathbf{j}$
   b. $\mathbf{F} = 2y\,\mathbf{i} + 3x\,\mathbf{k}$
   c. $\mathbf{F} = 5z\,\mathbf{i} + 2y\,\mathbf{j} - 7x\,\mathbf{k}$

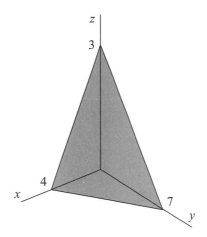

FIGURE 3.33. Surface for Exercise 3.15.

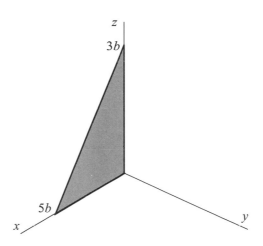

FIGURE 3.34. Surface for Exercise 3.16.

16. Verify Stokes's theorem for each of the following vector fields $\mathbf{F}$ using the plane surface shown in Fig. 3.34.

a. $\mathbf{F} = z^2\,\mathbf{i} + x^2\,\mathbf{j}$
b. $\mathbf{F} = -2z^3\,\mathbf{i}$
c. $\mathbf{F} = 2x^2\,\mathbf{i} + 3xz\,\mathbf{j} + (x^2 + 2xz)\,\mathbf{k}$

17. Verify Stokes's theorem for each of the following vector fields $\mathbf{F}$ using the plane surface shown in Fig. 3.35.

a. $\mathbf{F} = y\,\mathbf{i} + 2z\,\mathbf{j} - 3x\,\mathbf{k}$
b. $\mathbf{F} = -y\,\mathbf{i} - x\,\mathbf{k}$
c. $\mathbf{F} = 5z\,\mathbf{i} + 3z\,\mathbf{j} + 2z\,\mathbf{k}$

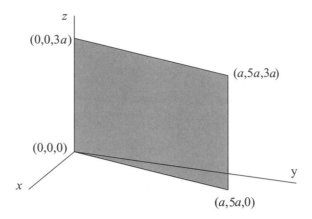

FIGURE 3.35. Surface for Exercise 3.17.

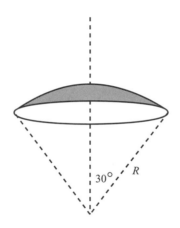

FIGURE 3.36. Surface for Exercises 3.19 and 3.20.

18. Check Stokes's theorem for each of the following vector fields using a hemispherical surface of radius R with the center of curvature located at the origin of the coordinate system and $z \geq 0$.

   a. $\mathbf{F} = \hat{\mathbf{r}} (r \cos^2 \theta) + \hat{\boldsymbol{\theta}} (r \cos \theta \sin \theta) + \hat{\boldsymbol{\phi}} (r \sin \theta)$
   b. $\mathbf{F} = \hat{\mathbf{r}} (r \sin \phi \cos \phi) + \hat{\boldsymbol{\theta}} (r \sin \theta \sin \phi) + \hat{\boldsymbol{\phi}} (r \sin^2 \theta)$

19. Check Stokes's theorem for the vector field

$$\mathbf{F} = \hat{\mathbf{r}} r \sin \theta \cos^2 \phi + \hat{\boldsymbol{\theta}} r \cos^2 \theta + \hat{\boldsymbol{\phi}} r \sin^2 \theta$$

   using the spherical cap shown in Fig. 3.36.

20. Check Stokes's theorem for the vector field

$$\mathbf{F} = \hat{\mathbf{r}}(r^2 \cos^2 \theta \cos^2 \phi) + \hat{\boldsymbol{\theta}}(r^2 \cos^2 \theta \sin^2 \phi) + \hat{\boldsymbol{\phi}}(r^2 \sin^2 \theta)$$

   using the spherical cap in Fig. 3.36.

21. In a region in which no electric charge exists, we see from Eq. (3.20) that $\nabla \cdot \mathbf{E} = 0$ everywhere in the region. We shall also show later[12] that an electrostatic field $\mathbf{E}$ satisfies the relation $\nabla \times \mathbf{E} = 0$. Suppose we have an electrostatic field that is parallel to the z-axis everywhere in a given region of space that contains no electric charge. Show that $\mathbf{E}$ must be a constant vector. Since the choice of coordinate system is arbitrary, what is the physical implication of this result?

22. From Eqs. (3.10) and (3.19), it is clear that the magnetic flux density $\mathbf{B}$ can be expressed in terms of a vector potential function $\mathbf{A}(\mathbf{r})$ according to

$$\mathbf{B}(\mathbf{r}) \equiv \nabla \times \mathbf{A}(\mathbf{r}).$$

A small magnetic dipole with magnetic dipole moment $\mathbf{m}$ produces a magnetic field in the space around it. At any point P far from this magnetic dipole (see Fig. 3.37), the vector potential takes the form

$$\mathbf{A}(\mathbf{r}) = k' \frac{\mathbf{m} \times \mathbf{r}}{r^3},$$

where k' is the magnetic force constant of Eq. (3.23). Use rectangular coordinates to show that the magnetic field far from a small magnetic dipole is given by

$$\mathbf{B}(\mathbf{r}) = k' \left[\frac{3(\mathbf{m} \cdot \mathbf{r})\mathbf{r}}{r^5} - \frac{\mathbf{m}}{r^3} \right].$$

23. Show that

$$(\mathbf{a} \times \mathbf{r})^2 = \sum_{i=1}^{3} \sum_{j=1}^{3} a_i a_j \left[\delta_{ij} \sum_{k=1}^{3} x_k^2 - x_i x_j \right],$$

where $\mathbf{a}$ is an arbitrary vector, $\mathbf{r} \equiv \mathbf{i}\,x_1 + \mathbf{j}\,x_2 + \mathbf{k}\,x_3$, and the symbol δ_{ij} is defined by

$$\delta_{ij} = \begin{cases} 1 & \text{for } i = j \\ 0 & \text{for } i \neq j. \end{cases}$$

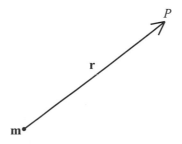

FIGURE 3.37. Magnetic dipole $\mathbf{m}$ for Exercise 3.22.

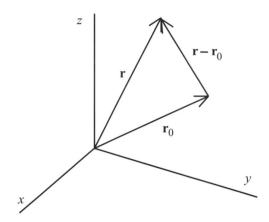

FIGURE 3.38. Position vectors for Exercise 3.24.

24. In rectangular coordinates, the difference between a variable position vector $\mathbf{r}$ and a constant position vector $\mathbf{r}_0$ is,

$$\mathbf{r} - \mathbf{r}_0 = \mathbf{i}\,(x - x_0) + \mathbf{j}\,(y - y_0) + \mathbf{k}\,(z - z_0).$$

This relationship is represented graphically in Fig. 3.38. Show that

$$\nabla\left(\frac{1}{|\mathbf{r} - \mathbf{r}_0|}\right) = -\frac{\mathbf{r} - \mathbf{r}_0}{|\mathbf{r} - \mathbf{r}_0|^3} \quad \text{for } \mathbf{r} \neq \mathbf{r}_0$$

and that

$$\nabla^2\left(\frac{1}{|\mathbf{r} - \mathbf{r}_0|}\right) = 0 \quad \text{for } \mathbf{r} \neq \mathbf{r}_0.$$

Define a vector $\mathbf{s} = \mathbf{r} - \mathbf{r}_0$ (hence, $\nabla \equiv \nabla_\mathbf{r} = \nabla_\mathbf{s}$) and use the divergence theorem to show that

$$\int_V \nabla^2\left(\frac{1}{|\mathbf{r} - \mathbf{r}_0|}\right) d^3s = -4\pi,$$

where the point at $\mathbf{r}_0$ lies inside the volume V.

HINT: For V, choose a sphere centered on the point at $\mathbf{r}_0$.

25. Maxwell's equations[13] are coupled equations in the electric and magnetic fields $\mathbf{E}$ and $\mathbf{B}$, respectively. For electromagnetic fields in free space, these equations can be written

$$\nabla \cdot \mathbf{E} = 0 \qquad \nabla \times \mathbf{E} + \frac{\partial \mathbf{B}}{\partial t} = 0$$

$$\nabla \cdot \mathbf{B} = 0 \qquad \nabla \times \mathbf{B} - \frac{1}{c^2}\frac{\partial \mathbf{E}}{\partial t} = 0,$$

---

[13] See Appendix B.

where c is the speed of light in free space. Use the vector identities in Appendix A.1 to show that the equations for the fields can be decoupled according to

$$\nabla^2 \mathbf{E} - \frac{1}{c^2}\frac{\partial^2 \mathbf{E}}{\partial t^2} = 0 \qquad \text{and} \qquad \nabla^2 \mathbf{B} - \frac{1}{c^2}\frac{\partial^2 \mathbf{B}}{\partial t^2} = 0.$$

26. The lengths of the sides of a parallelogram are a and b. Represent these sides as vectors $\mathbf{a}$ and $\mathbf{b}$ and show that the area A of the parallelogram is given by

$$A = |\mathbf{a} \times \mathbf{b}|. \qquad \text{(parallelogram)}$$

With the tail of vector $\mathbf{a}$ touching the tail of vector $\mathbf{b}$, construct a triangle by drawing a straight line from the head of $\mathbf{a}$ to the head of $\mathbf{b}$. Show that the area A of this triangle is

$$A = \tfrac{1}{2}|\mathbf{a} \times \mathbf{b}|. \qquad \text{(triangle)}$$

27. A particle with electric charge e moves with velocity $\mathbf{v}$ in a magnetic field $\mathbf{B}$. The magnetic force on the particle is

$$\mathbf{F}_{\text{mag}} = e\mathbf{v} \times \mathbf{B}.$$

If we express the magnetic field at position $\mathbf{r}$ at time t in terms of a vector $\mathbf{A}(\mathbf{r}, t)$, according to

$$\mathbf{B}(\mathbf{r}, t) = \nabla \times \mathbf{A}(\mathbf{r}, t),$$

Eq. (3.10) assures that Eq. (3.19) is satisfied. Show by explicit differentiation in rectangular coordinates that the magnetic force on the particle can be written

$$\mathbf{F}_{\text{mag}}(\mathbf{r}, t) = e\big[\nabla(\mathbf{v} \cdot \mathbf{A}) - (\mathbf{v} \cdot \nabla)\mathbf{A}\big].$$

28. The vectors $\mathbf{F}$ and $\mathbf{G}$ are arbitrary functions of position $\mathbf{r}$. Starting with the relations $\mathbf{F} \times (\nabla \times \mathbf{G})$ and $\mathbf{G} \times (\nabla \times \mathbf{F})$, obtain the identity

$$\nabla(\mathbf{F} \cdot \mathbf{G}) = (\mathbf{F} \cdot \nabla)\mathbf{G} + (\mathbf{G} \cdot \nabla)\mathbf{F} + \mathbf{F} \times (\nabla \times \mathbf{G}) + \mathbf{G} \times (\nabla \times \mathbf{F}).$$

29. For any two vector functions $\mathbf{F}(\mathbf{r})$ and $\mathbf{G}(\mathbf{r})$, use rectangular coordinates to establish the identity,

$$\nabla \times (\mathbf{F} \times \mathbf{G}) = \mathbf{F}(\nabla \cdot \mathbf{G}) + (\mathbf{G} \cdot \nabla)\mathbf{F} - \mathbf{G}(\nabla \cdot \mathbf{F}) - (\mathbf{F} \cdot \nabla)\mathbf{G}.$$

30. The spherical polar coordinate representation of the gradient operator is

$$\nabla = \hat{\mathbf{r}}\frac{\partial}{\partial r} + \hat{\boldsymbol{\theta}}\frac{1}{r}\frac{\partial}{\partial \theta} + \hat{\boldsymbol{\phi}}\,\frac{1}{r\sin\theta}\frac{\partial}{\partial \phi}.$$

In this coordinate system, an arbitrary vector $\mathbf{F}$ is represented in terms of its components by

$$\mathbf{F} = \hat{\mathbf{r}}\,F_r + \hat{\boldsymbol{\theta}}F_\theta + \hat{\boldsymbol{\phi}}\,F_\phi.$$

The rectangular components of the unit vectors $(\hat{\mathbf{r}}, \hat{\boldsymbol{\theta}}, \hat{\boldsymbol{\phi}})$ are

$$\hat{\mathbf{r}} = \hat{\mathbf{i}} \sin\theta \cos\phi + \hat{\mathbf{j}} \sin\theta \sin\phi + \hat{\mathbf{k}} \cos\theta,$$
$$\hat{\boldsymbol{\theta}} = \hat{\mathbf{i}} \cos\theta \cos\phi + \hat{\mathbf{j}} \cos\theta \sin\phi - \hat{\mathbf{k}} \sin\theta,$$
$$\hat{\boldsymbol{\phi}} = -\hat{\mathbf{i}} \sin\phi + \hat{\mathbf{j}} \cos\phi.$$

Verify that these are mutually perpendicular *unit* vectors and that their directions are geometrically consistent with Fig. 3.4. Note that in this representation the unit vectors $(\hat{\mathbf{r}}, \hat{\boldsymbol{\theta}}, \hat{\boldsymbol{\phi}})$ are functions of the polar coordinates θ and ϕ. Calculate the partial derivative of each of the unit vectors $(\hat{\mathbf{r}}, \hat{\boldsymbol{\theta}}, \hat{\boldsymbol{\phi}})$ with respect to each of the polar coordinates (r, θ, ϕ). Now take the scalar products of these derivatives with each of the unit vectors $(\hat{\mathbf{r}}, \hat{\boldsymbol{\theta}}, \hat{\boldsymbol{\phi}})$. Using the spherical polar representations of ∇ and $\mathbf{F}$ given above, construct $\nabla \cdot \mathbf{F}$, keeping in mind that the differential operators also operate on the unit vectors $(\hat{\mathbf{r}}, \hat{\boldsymbol{\theta}}, \hat{\boldsymbol{\phi}})$. (The unit vectors $(\mathbf{i}, \mathbf{j}, \mathbf{k})$, of course, are constant.) Show that your result can be cast in the form

$$\nabla \cdot \mathbf{F} = \frac{1}{r^2} \frac{\partial}{\partial r}(r^2 F_r) + \frac{1}{r\sin\theta} \frac{\partial}{\partial\theta}(\sin\theta\, F_\theta) + \frac{1}{r\sin\theta} \frac{\partial F_\phi}{\partial\phi}.$$

4

Complex Numbers

4.1   Why Study Complex Numbers?

Complex numbers play a very useful role in applied mathematics. As an example, let us consider the behavior of a particle of mass m constrained to move in one dimension under the influence of an elastic force,

$$F = -kx$$

where x is the displacement of the particle from its equilibrium position and k is a constant. This relation is known as *Hooke's law*. From Newton's second law of motion ($F = ma$), we have

$$-kx = m \frac{d^2 x}{dt^2}$$

or

$$\frac{d^2 x}{dt^2} + \frac{k}{m} x = 0. \tag{4.1}$$

This is a very simple differential equation which we can solve by inspection.[1]

From Eq. (4.1), we see that x must be such that when we differentiate it twice with respect to t, we obtain x multiplied by a constant. An exponential function

---

[1]In section 5.3, we shall solve this equation in a different way to demonstrate a more powerful method for obtaining solutions to linear differential equations.

has this property. If we substitute $x = e^{ct}$ into Eq. (4.1), we get

$$(c^2 + \frac{k}{m})x = 0$$

for all values of x. Thus,

$$c^2 = -\frac{k}{m} < 0.$$

There is no *real* number c which satisfies this quadratic equation. However, there are two roots to the equation which we can write as

$$c = \pm i \sqrt{\frac{k}{m}}, \qquad (4.2)$$

where we have denoted the positive square root of -1 by $i \equiv \sqrt{-1}$.

The general solution to Eq. (4.1) is then

$$x(t) = A e^{i\sqrt{k/m}\, t} + B e^{-i\sqrt{k/m}\, t}, \qquad (4.3)$$

where A and B are arbitrary constants of integration. The quantity c in Eq. (4.2) is said to be *imaginary*. In general, an imaginary number c may be written as $c = ib$ where b is real. The sum of a real and an imaginary number is said to be *complex*. That is, if a and b are both real, then the number

$$a + ib$$

is complex. Similarly, a complex *variable* z may be written as

$$z = x + iy$$

where the real variables x and y are called, respectively, the *real part of z* and the *imaginary part of z*. An alternate notation that is often used in the physics literature is $x = \text{Re}(z)$ and $y = \text{Im}(z)$.

It is useful to represent z as a point in a plane as illustrated in Fig. 4.1. This xy-plane is called the *complex plane*. From Fig. 4.1, we see that

$$x = r \cos \phi \qquad y = r \sin \phi$$

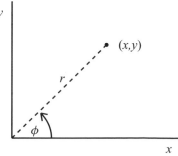

FIGURE 4.1. The complex plane.

and

$$r = \sqrt{x^2 + y^2} \qquad \phi = \tan^{-1} \frac{y}{x}.$$

In polar coordinates, we write

$$z = r(\cos \phi + i \sin \phi) = re^{i\phi},$$

where the second equality follows from[2]

$$e^{i\phi} = \cos \phi + i \sin \phi. \tag{4.4}$$

Example 1.
Write the complex number $\sqrt{1 + 2i}$ explicitly in terms of its real and imaginary parts and in its polar form $re^{i\phi}$.

Solution.
Set $\sqrt{1 + 2i} = a + ib$ and square both sides of this equation to get

$$1 + 2i = a^2 - b^2 + 2iab.$$

Comparing real and imaginary parts, we see that $a^2 - b^2 = 1$ and $ab = 1$. Solving these equations simultaneously for a and b and requiring that a and b be real, we find that

$$a = \sqrt{\frac{\sqrt{5} + 1}{2}} \qquad \text{and} \qquad b = \sqrt{\frac{\sqrt{5} - 1}{2}}.$$

From these results, we obtain

$$r = \sqrt{a^2 + b^2} = 5^{\frac{1}{4}} \qquad \text{and} \qquad \phi = \tan^{-1} \frac{b}{a} = \tan^{-1}\left(\frac{\sqrt{5} - 1}{2}\right).$$

Hence,

$$\sqrt{1 + 2i} = \sqrt{\frac{\sqrt{5} + 1}{2}} + i\sqrt{\frac{\sqrt{5} - 1}{2}} = 5^{\frac{1}{4}} e^{i \tan^{-1}\left(\frac{\sqrt{5}-1}{2}\right)}.$$

Example 2.
Write the complex number $c = 4e^{i\frac{\pi}{4}}$ explicitly in the form $c = \sqrt{a + ib}$ where a and b are exact real numbers.

Solution.
Using Eq. (4.4), we write

$$c = 4e^{i\frac{\pi}{4}} = 4(\cos \tfrac{\pi}{4} + i \sin \tfrac{\pi}{4}) = \tfrac{4}{\sqrt{2}}(1 + i) \equiv \sqrt{a + ib}.$$

To find a and b, we square both sides of this last equality to get

$$8(0 + 2i) = a + ib,$$

---

[2]This relation, known as *Euler's equation*, is derived in Section 5.2.

from which we see that $a = 0$ and $b = 16$. Thus,

$$c = 4e^{i\frac{\pi}{4}} = \sqrt{0 + 16i} = 4\sqrt{i}.$$

The *complex conjugate* of a complex variable $z = x + iy$ is denoted by z^* and is defined by

$$z^* = x - iy.$$

That is, the complex conjugate of z is obtained by replacing i by $-i$ in z.

The *absolute value* of z is defined by

$$|z| = \sqrt{z^*z}.$$

On evaluating the product $z^*z = (x - iy)(x + iy)$, we have

$$|z| = \sqrt{x^2 + y^2} = r.$$

A *function* $f(z)$ of a complex variable z is complex and can be written as a sum of its real and imaginary parts according to

$$f(z) = u(x, y) + iv(x, y) \qquad \text{with} \quad z = x + iy. \qquad (4.5)$$

Here u and v are real functions of the real variables x and y.

Example 3.

Write the function

$$f(z) = \frac{1}{z^2}$$

in terms of its real and imaginary parts $u(x, y)$ and $v(x, y)$ as indicated in Eq. (4.5) and in the form $re^{i\phi}$.

Solution.

Substitute $z = x + iy$ in the expression for $f(z)$, multiply the numerator and denominator by a factor that will make the denominator real, and identify the real and imaginary parts of the result to obtain

$$f(z) = u(x, y) + iv(x, y)$$
$$= \frac{x^2 - y^2}{(x^2 + y^2)^2} + i\frac{-2xy}{(x^2 + y^2)^2}.$$

From this result, we also find that

$$f(z) = |f(z)|e^{i\tan^{-1}(v/u)} = \frac{1}{x^2 + y^2}e^{i\tan^{-1}[-2xy/(x^2-y^2)]}$$

4.2   Roots of a Complex Number

In general, there are n values of z that satisfy the equation

$$z^n = a, \qquad (4.6)$$

where z and a may be complex. These values are called the *roots* of z^n or the nth roots of a. To find expressions for the roots in terms of the known quantity a, we write z and a in polar representation

$$z = |z|e^{i\theta} \tag{4.7}$$

$$a = |a|e^{i\phi} = |a|e^{i(\phi+2m\pi)}, \tag{4.8}$$

where m is any integer. Substituting these expressions in Eq. (4.6), we have

$$z^n = |z|^n e^{in\theta} = |a|e^{i(\phi+2m\pi)}.$$

The real part on the left-hand side of this equation must be equal to the real part on the right-hand side, with a similar equality for the imaginary parts. Thus,

$$|z| = |a|^{\frac{1}{n}} \tag{4.9}$$

$$e^{i\theta} = e^{i(\phi+2m\pi)/n} \qquad \text{for} \quad m = 0, 1, 2, \ldots n-1. \tag{4.10}$$

These equations provide values for $|z|$ and $\theta = (\phi + 2m\pi)/n$ in Eq. (4.7) in terms of the known quantities $|a|$ and ϕ given in Eq. (4.8).

Example 4.
Find the roots of

$$z^3 = 8.$$

Solution.
On comparing this equation with Eqs. (4.6) and (4.8), we identify $|a| = 8$ and $\phi = 0$. Thus, $|z| = 2$ and $\theta = 0, 2\pi/3$, and $4\pi/3$ from which we obtain the three roots

$$z = 2 \qquad \text{and} \qquad z = -1 \pm i\sqrt{3}.$$

Example 5.
Obtain the roots of

$$z^6 = -64.$$

Solution.
First, we write

$$z^6 = -2^6 = 2^6 e^{i\pi}.$$

Then, from Eqs. (4.9) and (4.10) we have

$$z = 2e^{i(2m+1)\pi/6}, \qquad \text{with} \quad m = 0, 1, 2, 3, 4, 5.$$

Explicitly, the six roots are

$$z = \pm 2i \qquad z = \sqrt{3} \pm i \qquad z = -\sqrt{3} \pm i.$$

4.3   Exercises

1. Write each of the following explicitly in the form

$$c = a + ib,$$

where a and b are real.

a. $c = i^3$

b. $c = i^5 + i^2$

c. $c = (2 - 3i) + (4 + 7i)$

d. $c = (1 + 2i)(2 - 3i)(4 + 5i)$

e. $c = (2 + 3i)/(4 - 5i)$

f. $c = \frac{4+3i}{3-2i} + \frac{4-6i}{5+i}$

g. $c = \left(\frac{3+i}{2+4i}\right)^2$

h. $c = \left(\frac{1+3i}{3+i}\right)^2 + \frac{5+2i}{4+i}$

2. Write each of the complex numbers c in Exercise 4.1 in the form $c = re^{i\phi}$ with explicit numerical expressions for r and ϕ.

3. Write each of the following complex functions explicitly in the form

$$f(z) = u(x, y) + iv(x, y),$$

where u and v are real functions of the real variables x and y with $z = x + iy$.

a. $f(z) = z^2$

b. $f(z) = z^3$

c. $f(z) = 1/(1 - z)$

d. $f(z) = z + z^{-1}$

e. $f(z) = 1/(1 - z^2)$

f. $f(z) = \sqrt{z}$

4. Write each of the complex functions $f(z)$ in Exercise 4.3 in the form

$$f(z) = re^{i\phi},$$

where r and ϕ are expressed explicitly in terms of x and y.

5. Using the relation in Eq. (4.4), the general solution to Eq. (4.1) given in Eq. (4.3) can also be written

$$x(t) = C \cos \sqrt{k/m}\, t + D \sin \sqrt{k/m}\, t,$$

where C and D are arbitrary constants. Express C and D explicitly in terms of the constants A and B of Eq. (4.3).

6. Show that each of the complex numbers c can be written explicitly in the form $c = \sqrt{a + ib}$ where a and b are exact real numbers.

a. $c = 3e^{i\pi/4}$

b. $c = \sqrt[4]{5}e^{i\,\tan^{-1}(\sqrt{5}-2)}$

c. $c = \sqrt[4]{8}e^{i\,\tan^{-1}\sqrt{(13-4\sqrt{10})/3}}$

d. $c = 13e^{i\,\tan^{-1}\frac{5}{12}}$

7. Obtain the explicit general solution $x(t)$ to the differential equation

$$\frac{d^2x(t)}{dt^2} - 3\frac{dx(t)}{dt} - \tfrac{7}{4}x(t) = 0,$$

by assuming $x(t) = e^{ct}$ where c is constant. Notice that this method of solution will work for any ordinary, linear differential equation in which the coefficients of the derivatives (including the zeroth) are all constants.[3]

8. Verify the relation

$$|e^{iz}| = e^{-\text{Im}(z)}.$$

9. By direct substitution, verify that the three roots obtained in Example 4.4 satisfy the equation

$$z^3 = 8.$$

10. Show by direct substitution that each of the six values obtained for z in Example 4.5 yields the result $z^6 = -64$.

11. Find all roots of each of the following:

$$z^4 = -81 \qquad\qquad z^4 = -8(1 - i\sqrt{3})$$
$$z^3 = i\,64 \qquad\qquad z^6 = -1 + i$$
$$z^4 = -i\,16 \qquad\qquad z^6 = -i$$

Express your results in the form $z = a + ib$. Check your answers by direct substitution for z in the original equation.

NOTE: In reducing your results to the required form, you may find the identities

$$\cos^2\theta = \frac{1 + \cos 2\theta}{2} \qquad \text{and} \qquad \sin^2\theta = \frac{1 - \cos 2\theta}{2}$$

useful.

---

[3]It is easily verified that in certain special cases, this substitution will yield only a single solution, as we shall see in Exercise 5.33.

5

Differential Equations

An ordinary differential equation expresses the relationship of a variable f (the *dependent* variable) to a second variable x (the *independent* variable). The equation contains terms involving derivatives of f with respect to x (i.e., $f'(x) = df/dx$, $f''(x) = d^2 f/dx^2$, etc.). For example, a linear, second-order differential equation has the general form

$$a(x)\frac{d^2 f(x)}{dx^2} + b(x)\frac{df(x)}{dx} + c(x)f(x) = s(x), \tag{5.1}$$

where a, b, c, and s are specific functions of x. The problem consists in solving this equation for f as an explicit function of x.

To illustrate the point and to put it in a familiar physical context, consider a particle of mass m falling freely near the earth's surface. From Newton's second law ($F = ma$), the equation of motion of the particle is

$$\frac{d^2 y}{dt^2} = g, \tag{5.2}$$

where g is the acceleration due to the earth's gravity and y is the position of the particle at time t. (We have chosen the direction toward the center of the earth to be positive.) In arriving at Eq. (5.2) from Newton's second law, we have made use of the fact that the velocity v of the particle is the time rate of change of the position ($v = dy/dt$) and the acceleration a is the time rate of change of the velocity ($a = dv/dt = d^2 y/dt^2$). Thus, Eq. (5.2) can be written as

$$\frac{dv}{dt} = g,$$

which we can integrate directly to get

$$v(t) = \int g \, dt = gt + A, \tag{5.3}$$

where A is an arbitrary constant of integration. From Eq. (5.3), we have

$$v = \frac{dy}{dt} = gt + A.$$

Now we integrate this equation to obtain y as a function of t,

$$y(t) = \int (gt + A) \, dt = \tfrac{1}{2}gt^2 + At + B,$$

where B is a second arbitrary constant of integration. Thus, the general solution to the differential equation Eq. (5.2),

$$y(t) = \tfrac{1}{2}gt^2 + At + B, \tag{5.4}$$

contains two arbitrary constants A and B. This is to be expected, because we had to integrate twice to find y as a function of t. Of course, integrating an equation like Eq. (5.1) will generally require a much more complicated procedure than the one we have used in this simple example, but the result will be similar. That is, the general solution to Eq. (5.1) will contain two arbitrary constants.

In practice, the arbitrary constants are fixed by physical constraints. For example, suppose that at time $t = 0$ the falling particle is at position y_0 and is falling with a velocity v_0. At time t, the particle has fallen to a new position y, which lies a distance $y - y_0$ below the point at y_0. On substituting $t = 0$ into Eqs. (5.3) and (5.4), we get $A = v_0$ and $B = y_0$, respectively. Thus, Eq. (5.4) becomes

$$y(t) = \tfrac{1}{2}gt^2 + v_0 t + y_0,$$

which is the familiar expression for the position at time t of a particle moving with constant acceleration g.

Note that any or all of the quantities a, b, c, and s in Eq. (5.1) may be constant or (with the exception of a) zero. Almost all of the ordinary differential equations we consider in this book will be of the general form in Eq. (5.1) with $s(x) = 0$, in which case the differential equation is said to be *homogeneous*.

Quite often in physics and applied mathematics the description of the behavior of a physical system is expressed as a differential equation, as we have seen here. Many of these equations of physical interest are *linear*. That is, whenever the dependent variable or one of its derivatives occurs in the equation, it appears in a given term with no other factors of the dependent variable or its derivatives. For example, Eq. (5.1) is linear because in each term on the left-hand side $f(x)$ appears only once, either as $f(x)$ itself or one of its derivatives. Equation (5.1) contains no terms like f^2, $f(df/dx)$, $(df/dx)^2$, $(df/dx)(d^2f/dx^2)$, $\ldots$. In this chapter, we shall consider a very powerful method for solving linear differential equations that involves the use of infinite series.

Nonlinear differential equations also arise in a variety of physical applications such as population growth, anharmonic oscillators, and chemical reactions. Treatment of these equations lies outside the scope of this book.

5.1 Infinite Series[1]

An infinite series is a sum containing an infinite number of terms with the kth term denoted by u_k We define the nth *partial sum* S_n by

$$S_n = u_1 + u_2 + \cdots + u_n = \sum_{k=1}^{n} u_k.$$

The series is said to converge and to have the value S, if the sequence of partial sums has the finite limit S, i.e.,

$$\lim_{n \to \infty} S_n - S.$$

It may happen that the terms u_k vary in sign. In this case, some terms with one sign will partly cancel the contributions of those with the opposite sign making the series more likely to converge. If the series

$$\sum_{k=1}^{\infty} |u_k|,$$

converges, then the series

$$\sum_{k=1}^{\infty} u_k$$

is said to *converge absolutely*. In any particular case, the convergence of a series may be established by one or more tests.

For the physical applications we want to consider, the terms in the infinite series will be functions $u_k(x)$ of a continuous variable x. In such cases, we are concerned with the convergence of

$$S(x) = \sum_{k=1}^{\infty} u_k(x) \qquad \text{for } a \leq x \leq b.$$

This series converges if, given any x in the interval $a \leq x \leq b$ and any $\epsilon > 0$, there exists an n such that the absolute difference between the series and the nth partial sum is less than ϵ, i.e.,

$$|S(x) - S_n(x)| < \epsilon \qquad \text{for } a \leq x \leq b.$$

---

[1]This discussion follows J. B. Seaborn, *Hypergeometric Functions and Their Applications*, Springer-Verlag, New York, 1991, p. 15. For a more complete treatment, see W. Kaplan, *op. cit.*, p. 310 or E. C. Titchmarsh, *The Theory of Functions*, Oxford University Press, Oxford, 1939.

Let n be the smallest integer for which this relation holds. In general, n depends on both ϵ and x. That is, $n = n(\epsilon, x)$.

If for each ϵ *another* number N (independent of x) can be found such that $n(\epsilon, x) < N(\epsilon)$ for all x in the interval, then the series $S(x)$ is said to *converge uniformly* in the interval. The virtues of uniform convergence include

- The sum of a uniformly convergent series of continuous functions is continuous.[2]
- A uniformly convergent series of continuous functions can be integrated term by term,

$$\int \left(\sum_{k=1}^{\infty} u_k(x) \right) dx = \sum_{k=1}^{\infty} \int u_k(x) \, dx.$$

- A uniformly convergent series of continuous functions can be differentiated term by term—*provided* the terms all have continuous derivatives and the series of derivatives converges uniformly.

5.2   Analytic Functions[3]

Generally, the mathematical functions of physics are *analytic*. We say that the function $f(x)$ is analytic for x in the interval $a < x < b$, if for every point x_0 in the interval we can write $f(x)$ as a convergent power series,

$$f(x) = \sum_{n=0}^{\infty} c_n(x - x_0)^n \tag{5.5}$$

where the coefficients c_n are numbers which are independent of x and x lies in the range $x_0 - R < x < x_0 + R$. The parameter $R \geq 0$ is called the *radius of convergence*. It is represented graphically in Fig. 5.1.

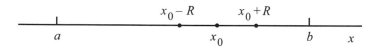

FIGURE 5.1. Radius of convergence.

By successive differentiations term by term in the series in Eq. (5.5), we obtain[4]

$$f'(x) = c_1 + 2c_2(x - x_0) + 3c_3(x - x_0)^2 + 4c_4(x - x_0)^3 + \cdots$$
$$f''(x) = 1 \cdot 2c_2 + 2 \cdot 3c_3(x - x_0) + 3 \cdot 4c_4(x - x_0)^2 + \cdots$$
$$f^{(k)}(x) = k!c_k + \frac{(k+1)!}{1!}c_{k+1}(x - x_0) + \frac{(k+2)!}{2!}c_{k+2}(x - x_0)^2 + \cdots.$$

On setting $x = x_0$ in this last expression for $f^{(k)}(x)$, we see that $f^{(k)}(x_0) = k! \, c_k$
or

$$c_k = \frac{1}{k!}\left[\frac{d^k}{dx^k} f(x)\right]_{x=x_0}.$$

Substitution of these results into Eq. (5.5) gives

$$f(x) = \sum_{k=0}^{\infty} \frac{f^{(k)}(x_0)}{k!}(x - x_0)^k. \tag{5.6}$$

This series is called the *Taylor series* of the function $f(x)$.

If a function $f(x)$ is analytic in the interval $a \leq x \leq b$, it can be expressed as a Taylor series expanded about any point x_0 in the interval. For expansion of the function about a given point, the Taylor series is unique.

Example 1.
Expand the function $f(x) = \sin x$ about the point $x = 0$.

Solution.
By direct calculation of the derivatives, we find that

$$c_k = \frac{1}{k!}\left[\frac{d^k}{dx^k} \sin x\right]_{x=0} = \begin{cases} \frac{(-1)^{(k-1)/2}}{k!} & \text{for } k \text{ odd} \\ 0 & \text{for } k \text{ even,} \end{cases}$$

from which we see that only odd powers of x are included in the Taylor series of $\sin x$. If we substitute these coefficients into Eq. (5.6) and set $k = 2n + 1$, then

$$\sin x = \sum_{n=0}^{\infty} \frac{(-1)^n x^{2n+1}}{(2n+1)!}, \tag{5.7}$$

where the sum is over *all* nonnegative integers.

Similarly, for $\cos x$ we obtain

$$\cos x = \sum_{n=0}^{\infty} \frac{(-1)^n x^{2n}}{(2n)!}. \tag{5.8}$$

---

[4]We shall often find it convenient to use the prime notation to denote first and second derivatives. That is, $y'(x) \equiv dy/dx$ and $y''(x) \equiv d^2y/dx^2$. For the kth derivative we may also write $y^{(k)}(x) \equiv d^k y/dx^k$.

The nth derivative of e^{ix} is easily found to be $i^n e^{ix}$. With this result in Eq. (5.6), we get the Taylor series of e^{ix} about $x = 0$,

$$e^{ix} = \sum_{n=0}^{\infty} \frac{(ix)^n}{n!}.$$

By summing this series for odd powers of x and even powers of x separately and comparing with Eqs. (5.7) and (5.8), we obtain the relation

$$e^{ix} = \cos x + i \sin x, \tag{5.9}$$

which is Euler's equation.

5.3   The Classical Harmonic Oscillator[5]

In Chapter 4, we considered the problem of a particle of mass m moving in one dimension under the influence of an elastic force given by Hooke's law,

$$F_{\text{elastic}} = -kx.$$

From Newton's second law of motion, we obtain the differential equation,

$$\frac{d^2x}{dt^2} + \omega^2 x = 0. \tag{5.10}$$

The parameter $\omega = \sqrt{k/m}$ is the natural angular frequency of the oscillator.

The problem is to solve this differential equation for position x as a function of time t. Let us try to find a solution which has the form

$$x(t) = \sum_{n=0}^{\infty} a_n t^{n+s}, \tag{5.11}$$

where s and the coefficients a_n are constants to be determined. We assume that this series converges uniformly for all values of t of physical interest. We substitute this series into the differential equation Eq. (5.10) and differentiate term by term. This gives

$$a_0 s(s-1)t^{s-2} + a_1(s+1)st^{s-1}$$
$$+ \sum_{n=0}^{\infty} [(n+s+2)(n+s+1)a_{n+2} + \omega^2 a_n]t^{n+s} = 0.$$

Terms corresponding to different powers of t are linearly independent. Since this equation holds for all values of t, we must have

$$s(s-1)a_0 = 0,$$
$$(s+1)sa_1 = 0, \tag{5.12}$$
$$(n+s+2)(n+s+1)a_{n+2} + \omega^2 a_n = 0.$$

---

[5] A similar treatment is given in Seaborn, *op. cit.*, p. 3.

If we take $s = 0$, then the first two of these equations are satisfied for arbitrary values of both a_0 and a_1. The third equation with $s = 0$ leads to

$$a_n = \frac{-\omega^2}{n(n-1)} a_{n-2} \qquad \text{for } n \geq 2.$$

Starting with a_0 (or a_1), we find by successive applications of this last equation that

$$a_n = \frac{(-1)^{\frac{n}{2}} \omega^n}{n!} a_0 \qquad n \text{ even,}$$

and

$$a_n = \frac{(-1)^{\frac{n-1}{2}} \omega^{n-1}}{n!} a_1 \qquad n \text{ odd.}$$

On substituting these coefficients into Eq. (5.11), we get

$$x(t) = a_0 \sum_{\substack{n=0 \\ n=\text{even}}}^{\infty} \frac{(-1)^{\frac{n}{2}} \omega^n t^n}{n!} + a_1 \sum_{\substack{n=1 \\ n=\text{odd}}}^{\infty} \frac{(-1)^{\frac{n-1}{2}} \omega^{n-1} t^n}{n!}.$$

In the first series with each value of n an even integer, we can redefine the summation index as k with $n \equiv 2k$ and k takes on all nonnegative integer values. Similarly, in the second sum where n is an odd integer, we define $n \equiv 2k + 1$ with k again running over all nonnegative integer values. With these definitions, we have

$$x(t) = a_0 \sum_{k=0}^{\infty} \frac{(-1)^k (\omega t)^{2k}}{(2k)!} + \frac{a_1}{\omega} \sum_{k=0}^{\infty} \frac{(-1)^k (\omega t)^{2k+1}}{(2k+1)!}$$

$$= a_0 \cos \omega t + \frac{a_1}{\omega} \sin \omega t.$$

Assume that at time $t = 0$ the particle is at $x = +b$ with a velocity $v = +v_0$. For these initial conditions, we find that

$$x(0) = a_0 = b$$

and

$$v(0) = \left[\frac{dx}{dt} \right]_{t=0} = a_1 = v_0.$$

The solution to the differential equation which satisfies the required initial conditions is seen to be

$$x(t) = b \cos \omega t + \frac{v_0}{\omega} \sin \omega t.$$

By making use of trigonometric identities, this solution may also be written as

$$x(t) = A \cos(\omega t + \phi)$$

where the constants A and ϕ are defined by

$$A = \sqrt{b^2 + \frac{v_0{}^2}{\omega^2}} \qquad \text{and} \qquad \phi = \tan^{-1}\left(\frac{-v_0}{b\omega}\right).$$

This is an example of a general approach to treating a problem in applied mathematics:

- Set up the differential equation describing the behavior of the system.
- Solve the equation (in this case by the series method).
- Invoke the boundary conditions to evaluate the integration constants.

5.4   Boundary Conditions

5.4.1   THE DAMPED OSCILLATOR

For another application of physical boundary conditions, consider the behavior of an elastically bound particle of mass m experiencing a viscous force given by

$$\mathbf{F}_{\text{viscous}} = -2m\gamma v,$$

where v is the velocity of the particle and γ is a constant. From Newton's second law, we obtain the equation of motion,

$$\frac{d^2x}{dt^2} + 2\gamma\frac{dx}{dt} + \omega_0^2 x = 0,$$

where $\omega_0 \equiv \sqrt{k/m}$. Inspection of this differential equation suggests that we look for solutions of the form $x = e^{ct}$. For this expression to satisfy the differential equation, we must put $c = -\gamma \pm \sqrt{\gamma^2 - \omega_0^2}$. So, for the general solution,

$$x(t) = Ae^{(-\gamma+\sqrt{\gamma^2-\omega_0^2})t} + Be^{(-\gamma-\sqrt{\gamma^2-\omega_0^2})t}.$$

To evaluate the constants A and B, let us assume that at time $t = 0$ the particle is released from rest at a distance b to the right of its equilibrium position as shown in Fig. 5.2.

First, we consider the case in which the viscous force is strong relative to the elastic force of the spring, i.e., $\gamma > \omega_0$. Setting $x(0) = b$ implies $A + B = b$. Similarly, on evaluating $v = dx/dt$ at $t = 0$ we get $A - B = \gamma b/\sqrt{\gamma^2 - \omega_0^2}$. We can solve these two equations simultaneously to obtain A and B. Hence, for these

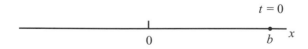

FIGURE 5.2. Harmonic oscillator displacement from equilibrium.

initial conditions, the position of the particle at time t is given by

$$x(t) = \frac{b}{2}e^{-\gamma t}\left[\left(1 + \frac{\gamma}{\sqrt{\gamma^2 - \omega_0^2}}\right)e^{\sqrt{\gamma^2 - \omega_0^2}\,t} + \left(1 - \frac{\gamma}{\sqrt{\gamma^2 - \omega_0^2}}\right)e^{-\sqrt{\gamma^2 - \omega_0^2}\,t}\right].$$

Since $\gamma > \omega_0$, both terms in $x(t)$ decay exponentially as functions of t. Physically, this means the viscous force is so strong the oscillator is damped out before it can complete a single oscillation.

Now, let us consider a viscous effect that is weaker than the elastic force provided by the spring, i.e., $\gamma < \omega_0$. For this case, it is convenient to define a real parameter $\omega_1 \equiv \sqrt{\omega_0^2 - \gamma^2}$ and write the solution to the equation of motion as

$$x(t) = e^{-\gamma t}\left[Ae^{i\omega_1 t} + Be^{-i\omega_1 t}\right].$$

Invoking the same initial conditions as before (i.e., the particle is released from rest at a distance b to the right of its equilibrium position), leads to an expression for the position of the particle at time t,

$$x(t) = be^{-\gamma t}\left[\cos \omega_1 t + \frac{\gamma}{\omega_1}\sin \omega_1 t\right].$$

Clearly, the particle oscillates with a frequency ω_1 related to the *natural* frequency ω_0 of the oscillator by $\omega_1 = \sqrt{\omega_0^2 - \gamma^2}$. So the particle does oscillate about its equilibrium position, but with an amplitude that steadily decreases because of the factor $e^{-\gamma t}$ arising from the viscous effect (i.e., $\gamma \neq 0$). Similarly, we obtain for the velocity of the particle at time t,

$$v(t) = -\omega_1 b\left(\frac{\gamma^2}{\omega_1^2} + 1\right)e^{-\gamma t}\sin \omega_1 t.$$

By substituting $t = 0$ into these expressions for $x(t)$ and $v(t)$, it is easily verified that they satisfy the initial conditions.

5.4.2   A Two-Dimensional Problem

A uniform electric field $\mathbf{E} = \mathbf{j}\,E$ exists in the region between two parallel, electrically charged plates separated by a distance d as illustrated in Fig. 5.3. A uniform magnetic field $\mathbf{B} = \mathbf{k}\,B$ is directed perpendicular to the electric field. A particle of mass m and positive electric charge $e > 0$ is released from rest at the bottom (positive) plate. The strengths of the fields are such that the particle barely grazes the upper plate. Our problem is to obtain equations for the trajectory of the particle. We ignore the effect of gravity and consider the electromagnetic force to be the only force on the particle.

The force the particle experiences is

$$\mathbf{F} = e(\mathbf{E} + \mathbf{v} \times \mathbf{B}).$$

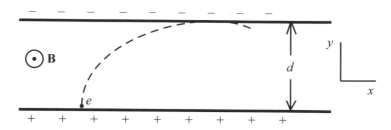

FIGURE 5.3. A charged particle in an electromagnetic field.

The rectangular components of this force are

$$F_x = ev_y B = eB\frac{dy}{dt},$$

$$F_y = e(E - v_x B) = e\left(E - B\frac{dx}{dt}\right).$$

From Newton's second law, we obtain the components of the equation of motion,

$$m\frac{d^2x}{dt^2} = eB\frac{dy}{dt}, \tag{5.13}$$

$$m\frac{d^2y}{dt^2} = e\left(E - B\frac{dx}{dt}\right). \tag{5.14}$$

For Eq. (5.13), we can write

$$\frac{d}{dt}\left(\frac{dx}{dt} - \frac{eB}{m}y\right) = 0.$$

Hence,

$$\frac{dx}{dt} = \frac{eB}{m}y + c, \tag{5.15}$$

where c is a constant. Substituting this expression in Eq. (5.14), we get

$$\frac{d^2y}{dt^2} = -\left(\frac{eB}{m}\right)^2\left[y + \frac{mc}{eB} - \frac{mE}{eB^2}\right]. \tag{5.16}$$

With the definition $u \equiv y + mc/eB - mE/eB^2$, Eq. (5.16) becomes

$$\frac{d^2u}{dt^2} + \left(\frac{eB}{m}\right)^2 u = 0.$$

From Section 5.3, we know the general solution to this equation to be

$$u(t) = a\cos\left(\frac{eBt}{m}\right) + b\sin\left(\frac{eBt}{m}\right) = y + \frac{mc}{eB} - \frac{mE}{eB^2},$$

or

$$y(t) = a\cos\left(\frac{eBt}{m}\right) + b\sin\left(\frac{eBt}{m}\right) - \frac{mc}{eB} + \frac{mE}{eB^2}.$$

This equation for $y(t)$ contains three arbitrary constants $a, b,$ and c, which are fixed by the boundary conditions,

- $y(0) = 0,$
- $v_y(0) = \dfrac{dy}{dt}\Big|_{t=0} = 0,$
- $v_x(0) = \dfrac{dx}{dt}\Big|_{t=0} = 0,$
- $v_y(t_1) = \dfrac{dy}{dt}\Big|_{t=t_1} = 0,$

where t_1 is the time at which the particle is at the top plate, i.e., $y(t_1) = d$. From the first two of these boundary conditions, we find that $b = 0, c = eBa/m + E/B$, and

$$y(t) = -a\left(1 - \cos\left(\frac{eBt}{m}\right)\right). \tag{5.17}$$

Applying the fourth boundary condition,

$$v_y(t_1) = \frac{dy}{dt}\Big|_{t=t_1} = -\frac{eBa}{m}\sin\left(\frac{eBt_1}{m}\right) = 0,$$

requires

$$t_1 = \frac{m\pi}{eB}n \qquad n = 0, 1, 2, \ldots.$$

Substituting t_1 for t in Eq. (5.17), we see that

$$y\left(\frac{m\pi}{eB}n\right) = 0, \qquad\qquad \text{for } n \text{ even,}$$

$$y\left(\frac{m\pi}{eB}n\right) = -2a = d \qquad \text{for } n \text{ odd.}$$

The first of these equations is consistent with the first boundary condition. The second corresponds to the position at time t_1 determined by the fourth boundary condition and gives $a = -d/2$. Thus, for the constant c we get

$$c = \frac{E}{B} - \frac{eBd}{2m}. \tag{5.18}$$

Hence,

$$y(t) = \frac{d}{2}\left(1 - \cos\left(\frac{eBt}{m}\right)\right). \tag{5.19}$$

Substituting from Eq. (5.19) for y and from Eq. (5.18) for c, Eq. (5.15) becomes

$$\frac{dx}{dt} = \frac{E}{B} - \left(\frac{eBd}{2m}\right)\cos\left(\frac{eBt}{m}\right),$$

which we integrate directly to obtain

$$x(t) = \frac{E}{B}t - \frac{d}{2}\sin\left(\frac{eBt}{m}\right). \tag{5.20}$$

In arriving at Eq. (5.20), we have chosen the integration constant such that $x(0) = 0$.

The third boundary condition yields a relation connecting the parameters E, B, and d,

$$v_x(0) = \frac{dx}{dt}\Big|_{t=0} = \frac{E}{B} - \frac{eBd}{2m}\cos\left(\frac{eBt}{m}\right)\Big|_{t=0} = \frac{E}{B} - \frac{eBd}{2m} = 0.$$

Using this result to eliminate d in Eqs. (5.19) and (5.20), we can write $x(t)$ and $y(t)$ exclusively in terms of the fields E and B,

$$x(t) = \frac{E}{B}\left[t - \frac{m}{eB}\sin\left(\frac{eBt}{m}\right)\right] \tag{5.21}$$

$$y(t) = \frac{mE}{eB^2}\left[1 - \cos\left(\frac{eBt}{m}\right)\right]. \tag{5.22}$$

The pair of equations, (5.21) and (5.22) provides a parametric representation of the path of the particle. It is clear that the particle oscillates between the two plates with a trajectory that is periodic. The repeating trajectory translates parallel to the plates.[6]

5.5   Polynomial Solutions

As the sequel will show, in certain applications physical constraints may dictate that in order to obtain a physically acceptable solution to a differential equation, the series must terminate, that is, the solution must be a polynomial (perhaps multiplied by some power of x, say x^s, where in the general case s is not necessarily an integer). To illustrate this point, we consider the differential equation

$$(1 - x^2)u''(x) - xu'(x) + \lambda u(x) = 0, \tag{5.23}$$

where λ is a real, nonnegative constant. Suppose we require that the solution be a polynomial of degree n. Find the constraint that this imposes on the parameter λ. Assuming a series solution of the form,

$$u(x) = \sum_{k=0}^{\infty} a_k x^{k+s}, \tag{5.24}$$

and following a procedure similar to that leading to Eqs. (5.12), we obtain from Eq. (5.23)

$$s(s-1)a_0 x^{s-2} + (s+1)sa_1 x^{s-1}$$
$$+ \sum_{k=0}^{\infty}\left\{(k+s+2)(k+s+1)a_{k+2} - \left[(k+s)^2 - \lambda\right]a_k\right\}x^{k+s} = 0.$$

---

[6]In Chapter 10, we shall encounter a particle moving along a similar path in a different physical context.

This equation must hold for all values of x in some continuous range. Hence,

$$s(s-1)a_0 = 0, \tag{5.25}$$

$$(s+1)sa_1 = 0, \tag{5.26}$$

$$(k+s+2)(k+s+1)a_{k+2} - \left[(k+s)^2 - \lambda\right]a_k = 0. \tag{5.27}$$

Given a_0, we can use Eq. (5.27) to obtain a_2. Using the result for a_2, Eq. (5.27) yields a_4, and so on for all a_k with k equal to an even integer. Similarly, starting with a_1, we can, by successive applications of Eq. (5.27), generate all *odd-indexed* coefficients a_k. This leads to a general solution of Eq. (5.23) of the form

$$u(x) = a_0 u_0(x) + a_1 u_1(x),$$

where $u_0(x)$ consists of x^s multiplied by a series with only even powers of x and $u_1(x)$ contains x^s multiplied by a series with only odd powers of x. As required for a second-order differential equation, this solution clearly has two arbitrary constants of integration, a_0 and a_1. It is also clear from this equation that if a_0 and a_1 are both zero, we obtain the trivial solution

$$u(x) = 0 \quad \text{for all } x.$$

For a nontrivial solution to Eq. (5.23), either a_0 or a_1 (or both) must be different from zero. We see from Eqs. (5.25) and (5.26), that this requirement constrains the parameter s. We have the two cases,

$$a_0 \neq 0 \Rightarrow \begin{cases} s = 0 \Rightarrow a_1 \text{ is arbitrary} \\ s = 1 \Rightarrow a_1 = 0 \end{cases}$$

$$a_1 \neq 0 \Rightarrow \begin{cases} s = 0 \Rightarrow a_0 \text{ is arbitrary} \\ s = -1 \Rightarrow a_0 = 0. \end{cases}$$

Consider the case for $s = 1$. From Eq. (5.26), we must have $a_1 = 0$. Therefore, a nontrivial solution implies $a_0 \neq 0$ and Eq. (5.27) becomes

$$a_{k+2} = \frac{(k+1)^2 - \lambda}{(k+3)(k+2)} a_k, \quad \text{with } k \text{ even.}$$

Requiring that the series in Eq. (5.24) be a polynomial of degree n means that $a_{n+1} = 0$, but $a_{n-1} \neq 0$. That is,

$$a_{n+1} = \frac{n^2 - \lambda}{(n+2)(n+1)} a_{n-1} = 0.$$

Noting that $k = n - 1$ is an even integer, we see that n is an odd integer and the choice of $s = 1$ in Eqs. (5.25)-(5.27) results in a polynomial only if λ is equal to the square of an odd integer. The resulting solution will contain only *odd* powers of x. It is left as an exercise to verify that $s = -1$ yields a solution that is a polynomial in *even* powers of x. These results show that Eq. (5.23) has a solution that is a polynomial of degree n, if $\lambda = n^2$.

Example 2.

Use the series method to solve the differential equation

$$(1 - x^2)u''(x) - 3xu'(x) + 35u(x) = 0, \tag{5.28}$$

Assume a series solution of the form in Eq. (5.24). Choose the number s such that a nontrivial solution *must* contain only odd powers of x. Show that this solution is a polynomial and write it out explicitly.

Solution.

Substitution of Eq. (5.24) into Eq. (5.28) leads to equations analogous to Eqs. (5.25) - (5.27),

$$s(s - 1)a_0 = 0, \tag{5.29}$$

$$(s + 1)sa_1 = 0, \tag{5.30}$$

$$(k + s + 2)(k + s + 1)a_{k+2} - \big[(k + s)(k + s + 2) - 35\big]a_k = 0. \tag{5.31}$$

Requiring that the solution *must* contain only odd powers of x implies that a_0 and a_1 cannot both be arbitrary. Hence, $s \neq 0$. If we choose $s = -1$, then Eq. (5.29) requires that $a_0 = 0$ and a_1 is arbitrary. In this case, only odd-indexed coefficients a_k are different from zero, leading to a series with only *even* powers of x. Therefore, we must choose $s = 1$. Then, Eqs. (5.29) and (5.30) require a_1 to be zero and a_0 to be arbitrary. With $s = 1$, Eq. (5.31) can be written as

$$a_{k+2} = \frac{(k + 1)(k + 3) - 35}{(k + 3)(k + 2)} a_k. \tag{5.32}$$

It is clear that for *some* value of k, say $k = n$, the numerator on the right-hand side of Eq. (5.32) vanishes and $a_n \neq 0$, but $a_{n+2} = 0$. Hence, all even-indexed coefficients a_k with $k > n$ are equal to zero. In this case, the numerator on the right-hand side of Eq. (5.32) is $(n + 1)(n + 3) - 35 = 0$, from which we find $n = 4$. Thus, for Eq. (5.28) the series solution given by Eq. (5.24) with $s = 1$ is a fifth-degree polynomial in x. We start with a_0, and through successive applications of Eq. (5.32) generate all even-indexed coefficients a_k to obtain the explicit solution,

$$u(x) = a_0\big(x - \tfrac{16}{3}x^3 + \tfrac{16}{3}x^5\big),$$

where a_0 is an arbitrary constant.

5.6   Elementary Functions

We can generalize the definition of the exponential function to complex z according to

$$e^z \equiv e^x(\cos y + i \sin y) = \mathrm{Re}(e^z) + i\,\mathrm{Im}(e^z), \tag{5.33}$$

where x and y are real variables. The real and imaginary parts of this function are analytic for all values of x and y, hence the function itself is analytic for all z.

For $y = 0$, we obtain the familiar (real) exponential e^x. On making use of Euler's formula, Eq. (5.9),

$$e^{iy} = \cos y + i \sin y,$$

we see that Eq. (5.33) can be written as

$$e^z = e^{x+iy}. \tag{5.34}$$

By analogy with real trigonometric and hyperbolic functions, we also have the following definitions

$$\cos z \equiv \frac{1}{2}\left[e^{iz} + e^{-iz}\right] = \cosh(iz)$$

$$\sin z \equiv \frac{1}{2i}\left[e^{iz} - e^{-iz}\right] = -i\sinh(iz)$$

$$\cosh z \equiv \frac{1}{2}\left[e^z + e^{-z}\right] = \cos(iz)$$

$$\sinh z \equiv \frac{1}{2}\left[e^z - e^{-z}\right] = -i\sin(iz).$$

5.7 Singularities[7]

Quite often the differential equations one encounters in theoretical physics are *linear, second-order, homogeneous* differential equations which have the general form

$$\frac{d^2}{dz^2}u(z) + P(z)\frac{d}{dz}u(z) + Q(z)u(z) = 0. \tag{5.35}$$

We shall refer to the form of the differential equation written in this way as the *standard form*. Most commonly the functions $P(z)$ and $Q(z)$ are *rational* functions of z.

The problem here is to find u as a function of z. Solving Eq. (5.35) algebraically for the second derivative $u''(x)$, we have

$$u''(z) = f(z, u, u'). \tag{5.36}$$

From this expression, we see that if u and u' are given at any point z_0, then $u''(z_0)$ can be obtained directly from Eq. (5.36). Differentiating Eq. (5.36) gives

$$\frac{d^3}{dz^3}u(z) = f'(z, u, u').$$

Since u and u' are known at z_0, we can also evaluate this expression at that point. If $u(z)$ is analytic at z_0, then all higher derivatives must exist at z_0 and we can write

---

[7] The discussion follows Seaborn, *op. cit.* p. 24.

the Taylor series for this function according to

$$u(z) = \sum_{n=0}^{\infty} \frac{u^{(n)}(z_0)}{n!}(z - z_0)^n.$$

If this series converges everywhere in a finite interval around z_0, then the solution exists for all values of z in this interval.

If u and u' can have *any* values at z_0, then z_0 is an *ordinary point* of the differential equation. If u and u' cannot be chosen arbitrarily at this point, then z_0 is a *singular point* of the differential equation. For example, consider the differential equation,

$$z^2 u''(z) + \lambda z u'(z) + \mu u(z) = 0,$$

where λ and μ are constants. Now set $z = 0$. We see from the differential equation that if $u(0) \neq 0$, then either $u'(0)$ or $u''(0)$ is *infinite*. Therefore, it is not possible to construct a Taylor series for this function for an interval around $z = 0$.

5.7.1   SINGULARITIES OF A FUNCTION

If the function $f(z)$ is analytic at a point $z = z_0$, then $f(z)$ is said to be *regular* at z_0. If $f(z)$ is *not* analytic at z_0, then it is *irregular* or *singular* at that point.

5.7.2   SINGULARITIES OF A DIFFERENTIAL EQUATION

We classify singular points of the differential equation, Eq. (5.35), in the following way: If $P(z)$ or $Q(z)$ is not analytic at z_0 so that $u''(z_0)$ cannot be evaluated to construct a Taylor series for $u(z)$ about that point, then the differential equation has a singularity at z_0. It is a *regular* singularity if and only if both

$$(z - z_0)P(z) \qquad \text{and} \qquad (z - z_0)^2 Q(z)$$

are analytic at z_0. Otherwise, the singularity at $z = z_0$ is *irregular*. Do not confuse the terms regular and irregular singular points of a differential equation with our use of the words *regular* and *irregular* in Section 5.7.1 to describe the behavior of a *function* at a given point.

At a singular point of Eq. (5.35), the series method may yield a solution that is singular at that point. Even in this case, an acceptable solution may still be obtained by imposing constraints on the series, as we shall see.[8]

5.8   Exercises

1. A particle of mass m, constrained to move horizontally, is bound by an elastic force given by

$$F = -m\omega^2 x,$$

---

[8] An example will be seen in Section 7.7.

where x is the position of the particle relative to its equilibrium position and ω is a real, positive constant. Use Newton's second law to obtain the equation of motion for this particle. Rewrite your equation in the standard form. The general solution to this equation can be written as an explicit function of time t according to

$$x(t) = A \sin \omega t + B \cos \omega t,$$

where A and B are arbitrary constants. At time $t = 0$ the particle is located at a distance b to the right of its equilibrium position and is moving to the left with speed v_0. Find the particle's position $x(t)$ and velocity $v(t)$ as explicit functions of time. Pay particular attention to signs. Express your results in terms of the given parameters b, v_0, and ω only. Next, write the position as a function of time in the form

$$x(t) = C \cos(\omega t + \phi),$$

where C and ϕ are constants expressed in terms of the given parameters. Explain clearly in words the physical meaning of the constant C.

2. A particle of mass m moves in one dimension under the influence of an elastic force F_{el} and a velocity-dependent viscous force F_{vis} given by

$$F_{el} = -m\omega_0^2 x \qquad \text{and} \qquad F_{vis} = -2m\gamma v,$$

where ω_0 and γ are real, positive constants with $\omega_0 > \gamma$ and x is the position of the particle relative to its equilibrium position. Write down the equation of motion (Newton's second law) for this particle and reduce it to a differential equation in standard form with x appearing explicitly as the dependent variable and the time t as the independent variable. The general solution to this differential equation can be written as

$$x(t) = e^{-\gamma t}\left[A \cos \sqrt{\omega_0^2 - \gamma^2}\, t + B \sin \sqrt{\omega_0^2 - \gamma^2}\, t\right],$$

where A and B are arbitrary constants. At time $t = 0$, the particle moving to the right with speed v_0 passes through the equilibrium position ($x = 0$). Find the position $x(t)$ as an explicit function of t. Obtain an expression for the maximum displacement of the particle from its equilibrium position after it passes through this position. Show graphically how x behaves as a function of t. Express all results only in terms of the given parameters γ, v_0, and ω_0.

3. The equation of motion for a damped harmonic oscillator can be written

$$\frac{d^2x(t)}{dt^2} + 2\gamma \frac{dx(t)}{dt} + \omega_0^2 x(t) = 0,$$

where γ and ω_0 are real, positive constants. For a weak viscous force ($\gamma < \omega_0$), the solution to this equation may be written in the form

$$x(t) = Ce^{-\gamma t} \sin\left(\sqrt{\omega_0^2 - \gamma^2}\, t + \phi\right).$$

At time $t = 0$, the particle is at $x = b$ and is moving with velocity v_0. Find the values of the constants C and ϕ such that these initial conditions are satisfied. Write out explicitly the corresponding expression for $x(t)$.

4. In general, the particle described in Section 5.4.2 has velocity components v_x and v_y parallel to and perpendicular to the plates, respectively. Find the maximum value of v_x. Where is the particle located relative to the plates when v_x has this value? What is the value of v_y when v_x has this value? Calculate the maximum *speed* of the particle on its trajectory. Where is the particle located relative to the plates when it is moving the fastest? What are the values of v_x and v_y when the particle is moving the fastest? Where is the particle located relative to the plates when v_y is maximum? What is v_x at this point? How fast is the particle moving at this point? Express all answers in terms of the given parameters e, m, B, and E.

5. Construct the Taylor expansion of $\cos x$ about $x = 0$ (i.e., verify Eq. (5.8)).

6. Set $\lambda = 49$ in Eq. (5.23) and obtain the explicit polynomial solution for $u(x)$ in this equation.

7. The equation

$$(1 - x^2)\frac{d^2u}{dx^2} - 2x\frac{du}{dx} + 12u = 0,$$

is a second-order, linear, homogeneous differential equation. Find the singular points of this differential equation. Determine whether each is regular or irregular. Explain your reasoning clearly in arriving at your answer in each case. Assume a series solution of the form,

$$u(x) = \sum_{n=0}^{\infty} a_n x^{n+s}$$

and choose a value for s such that your solution contains only even powers of x. For a different value of s, obtain a solution containing only odd powers of x. Show that one of these solutions is a polynomial.

8. Show that for $s = -1$ in Eq. (5.24) the series solution to Eq. (5.23) is a polynomial in x if λ is equal to the square of an *even* integer. Show also that this polynomial contains only even powers of x.

9. A solution to the differential equation

$$\rho\frac{d^2u(\rho)}{d\rho^2} + \frac{du(\rho)}{d\rho} + \rho u(\rho) = 0$$

has the form

$$u(\rho) = \sum_{k=0}^{\infty} a_k \rho^{k+s}$$

where s is a number. Obtain the general recursion formula for the coefficients a_k. Can both a_0 and a_1 be zero at the same time? Explain. What values may s have? Under what conditions? For each value of s, obtain the explicit expression for the coefficient a_k. Express your result in terms of factorials. Write down the explicit series solution corresponding to each value of s. Show that the two solutions are not independent by showing that one reduces to the other (except possibly for an overall multiplicative constant).

10. Use the series method to obtain the general solution to the differential equation

$$(1 - x^2)\frac{d^2 y}{dx^2} - 3x\frac{dy}{dx} + 24y = 0.$$

One of the solutions is a polynomial. Write down the polynomial solution explicitly showing all terms.

11. Find the singular points of the differential equation

$$x\frac{d^2 f}{dx^2} + (1 - x)\frac{df}{dx} + 4f = 0$$

and determine whether each is regular or irregular. Explain clearly and completely your reasoning in each case. Assume a series solution of the form

$$f(x) = \sum_{k=0}^{\infty} a_k x^{k+s}$$

and obtain the recursion formula for the coefficients a_k. What are the constraints on the coefficient a_0? What other implication does this have for the solution? Explain your reasoning clearly in each case. Write down explicitly the corresponding series expression for $f(x)$. Show that this series reduces to a polynomial. Write down the polynomial solution explicitly with exact values for the coefficients of the powers of x (except for an overall multiplicative constant).

12. Use the series method to obtain a solution to the differential equation

$$t^2\frac{d^2 g(t)}{dt^2} + t\frac{dg(t)}{dt} + (t^2 - 1)g(t) = 0.$$

13. Use the series method to find the general solution to the differential equation,

$$(1 - z^2)\frac{d^2 f}{dz^2} - z\frac{df}{dz} - \lambda f = 0,$$

where λ is a constant. Find the value of the parameter λ for which a solution to this differential equation is a sixth-degree polynomial. Write down this polynomial explicitly (except for an overall multiplicative constant).

14. For finite values of t, find the singular points of the differential equation

$$t\frac{d^2 x}{dt^2} + (4 - t)\frac{dx}{dt} + 5x = 0.$$

Determine whether each is regular or irregular. Assume a series solution of the form Eq. (5.11) and obtain the recursion formula for the coefficients a_n. What are the constraints on a_0? Explain. Find the series solution that is regular at $t = 0$. Show clearly that this solution is a polynomial. Construct the polynomial solution with *exact* coefficients of the powers of t (except for an overall multiplicative constant).

15. Use the series method to obtain solutions to the differential equation

$$x\frac{d^2u}{dx^2} + (\frac{3}{2} - x)\frac{du}{dx} + 4u = 0.$$

Find the solution that is regular at $x = 0$ and show that it is a polynomial. Write down the polynomial with *exact* coefficients of the powers of x (except for an overall multiplicative constant).

16. Obtain the explicit polynomial solution to the differential equation

$$(1 - x^2)u''(x) + \frac{7}{2}xu'(x) + \frac{5}{2}u(x) = 0.$$

17. Solve the differential equation

$$xu''(x) + xu'(x) - 4u(x) = 0,$$

using the series method. Show that this method yields only one solution. Show that this solution is a polynomial and write down the polynomial explicitly.

18. Obtain the recursion formula for the coefficients of the powers of x in the series solution of the differential equation,

$$x(1 - x)u''(x) + \lambda u'(x) + \mu u(x) = 0,$$

where λ and μ are constant parameters independent of x with $\lambda > 0$. For $\lambda = \frac{3}{2}$, find two (different) values of μ for which $u(x)$ is equal to a simple function of x multiplied by a polynomial of degree six. For each of these, write out the expression for $u(x)$ explicitly.

19. Find the constraint on the parameter $\alpha > 0$ such that the series solution to the differential equation,

$$(x - L)u''(x) - \alpha u'(x) = 0,$$

is a polynomial of degree n. In this equation, L is a constant. Find a closed expression for the general coefficient c_k in the sum,

$$u(x) = \sum_{k=0}^{\infty} c_k x^k,$$

and write down this sum with your expression substituted for c_k. Do you recognize this sum?

HINT: Write down the explicit expressions for $u(x)$ for $n = 3$ and $n = 4$.

20. Solve the differential equation

$$x(1 - x)\frac{d^2y(x)}{dx^2} - 4x\frac{dy(x)}{dx} + 40y(x) = 0$$

using the series method assuming a solution of the form

$$y(x) = \sum_{k=0}^{\infty} a_k x^{k+s}.$$

Obtain the recursion formula for generating the coefficients a_k. State clearly the constraints on a_0. Show that this method yields only one solution to this differential equation and that it is a polynomial. Write down the polynomial with exact coefficients of the powers of x.

21. The differential equation

$$\frac{d^2u(x)}{dx^2} + 2\lambda x \frac{du(x)}{dx} + (\mu + \lambda^2 x^2)u(x) = 0,$$

contains two real, positive, constant parameters μ and λ with $\lambda > \mu$. Find the general solution to this differential equation by making the change of dependent variable $u(x) = f(x)g(x)$. Substitute this product into the differential equation and choose $g(x)$ so that the first derivative term does not appear in the differential equation for $f(x)$. Solve the resulting equation for $f(x)$ and write down the general solution to the original equation for $u(x)$.

22. Use the series method to obtain two solutions to the differential equation

$$\frac{d^2u(x)}{dx^2} + 2x \frac{du(x)}{dx} + 2u(x) = 0.$$

Show clearly that one of these solutions is an elementary function.

23. The differential equation

$$x^2 u''(x) - 2x(1 - x)u'(x) + 2(1 - x)u(x) = 0$$

can be solved using the series method. The method yields two linearly independent solutions. From your results, show that the general solution to this differential equation can be written entirely in terms of elementary functions.

24. In the differential equation

$$\frac{d^2 f(x)}{dx^2} + 2x \frac{df(x)}{dx} + (x^2 + \mu)f(x) = 0$$

the parameter μ is a constant ($\mu > 1$). To solve this equation, make the change of variable

$$f(x) = e^{-\frac{1}{2}x^2} g(x)$$

and obtain a differential equation for $g(x)$. Solve the equation for $g(x)$ to get the general solution to the equation for $f(x)$ explicitly in terms of elementary functions. Find the specific expression for $f(x)$ that satisfies the two boundary conditions

$$f(0) = 0,$$
$$\frac{df(x)}{dx}\bigg|_{x=0} = \lambda,$$

where λ is a real, positive constant.

25. A motor car of mass m is driven along a straight, level roadway. The engine supplies a constant force F along the direction of motion. As the car moves along the roadway, it experiences a retarding frictional force proportional to its velocity v. That is, $F_{\text{friction}} - -m\gamma v$. Use Newton's second law to obtain

the equation of motion describing the behavior of the position x as a function of time t. Verify by substitution that your equation is satisfied by the function

$$x(t) = Ae^{-\gamma t} + \frac{F}{\gamma m}t + B,$$

where A and B are arbitrary constants. Find A and B such that the initial conditions

$$x(0) = b \qquad \text{and} \qquad v(0) = v_0$$

are met. Write down explicitly the corresponding expression for $x(t)$. Write down the Taylor series for $e^{-\gamma t}$ expanded about $t = 0$ showing explicitly the first four terms of this expansion. Substitute this result into your expression for $x(t)$, combining terms where possible. Now take the limit $\gamma \to 0$ and discuss your result. Is this result reasonable? Explain clearly.

26. A particle of mass m is confined to a horizontal plane. It is elastically bound to its equilibrium position by an isotropic elastic force,

$$\mathbf{F} = -m\omega^2 \mathbf{r},$$

where $\mathbf{r}(t)$ is the displacement of the particle from its equilibrium position and ω is a real, constant parameter. From Newton's law, we obtain the equation of motion for the particle,

$$\frac{d^2\mathbf{r}}{dt^2} + \omega^2\mathbf{r} = 0.$$

In rectangular coordinates, $\mathbf{r}(t) = \mathbf{i}\,x(t) + \mathbf{j}\,y(t)$. Obtain the expressions for $x(t)$ and $y(t)$ that satisfy the initial conditions on the position and velocity components,

$$x(0) = b \qquad y(0) = 0 \qquad v_x(0) = 0 \qquad v_y(0) = v_0.$$

Eliminate the time t from your equations to obtain the direct relationship between x and y. Based on your result, describe the shape of the path traced out by the particle as it moves in the plane of motion. Describe this path in the special case where $v_0 = b\omega$.

27. Solutions to the differential equation

$$\frac{d^2x(t)}{dt^2} + \lambda t \frac{dx(t)}{dt} + \mu x(t) = 0$$

may be obtained using the series method. How must the constant parameters λ and μ be related if one of the solutions $x(t)$ is a polynomial? If $\mu = \frac{1}{3}$, find the value of λ such that one of the solutions is a sixth-degree polynomial in t. Write out this polynomial explicitly.

28. The series method leads to two linearly independent solutions, $u_1(x)$ and $u_2(x)$, to the differential equation

$$\frac{d^2u(x)}{dx^2} + x\frac{du(x)}{dx} + u(x) = 0.$$

From these, construct the general solution,

$$u(x) = Au_1(x) + Bu_2(x).$$

Obtain values for the constants A and B such that the following boundary conditions are satisfied:

- $u(0) = 2$
- $u(\frac{1}{\sqrt{2}}) = 2e^{-\frac{1}{4}} - \sum_{n=0}^{\infty} \frac{(-1)^n n!}{(2n+1)!}$

Write explicitly the solution $u(x)$ that satisfies these boundary conditions.

29. In the differential equation,

$$x^2 \frac{d^2 f(x)}{dx^2} - 2x \frac{df(x)}{dx} + (2 - \mu^2 x^2) f(x) = 0,$$

the parameter μ is a real, positive constant. To obtain the general solution to this equation, assume a product solution of the form $f(x) = u(x)v(x)$ and substitute it into the differential equation. Choose $v(x)$ such that the first-order term in the differential equation for $u(x)$ vanishes for all x. Solve the resulting equation for $u(x)$ and construct the general solution to the differential equation for $f(x)$. Now solve the equation for $f(x)$ directly by the series method and show that the general solutions obtained by the two methods are equivalent.

30. A musket ball of mass m is fired vertically upward with an initial speed v_0. It rises to a maximum height h and then falls vertically downward. Throughout its trajectory, the ball experiences a viscous force proportional to its velocity v,

$$F_{\text{viscous}} = -m\gamma v,$$

where γ is a real, positive constant. During its rise, the ball has two downward forces exerted on it, F_{viscous} and the gravitational force mg where g is the acceleration due to the earth's gravity. Show that the equation of motion for this part of the ball's trajectory is

$$\frac{d^2 y}{dt^2} + \gamma \frac{dy}{dt} + g = 0,$$

where y is the position of the ball at time t. The solution to this differential equation is

$$y(t) = Ae^{-\gamma t} - \frac{g}{\gamma} t + B,$$

where A and B are arbitrary constants. Impose the initial conditions to evaluate A and B and obtain an expression for $y(t)$ that satisfies these initial conditions. Let t_1 be the time the ball reaches its maximum height h. Show that

$$t_1 = \frac{1}{\gamma} \log\left(1 + \frac{v_0 \gamma}{g}\right).$$

Use this result to find an expression for h in terms of the parameters g, γ, and v_0.

Now consider the ball falling a distance h back to its original height. Again the ball experiences a viscous force in the direction opposite to its motion. For this part of the trajectory, we restart the clock at $t = 0$ and take the downward direction to be positive. Show that the corresponding equation of motion is

$$\frac{d^2x}{dt^2} + \gamma \frac{dx}{dt} - g = 0,$$

where the position of the ball is now denoted by x. The general solution to this equation is

$$x(t) = Ce^{-\gamma t} + \frac{g}{\gamma}t + D,$$

where C and D are arbitrary constants. Let t_2 denote the time the ball reaches the height from which it was first fired and calculate $x(t_2)$ from the last expression. Show that the velocity at this point is

$$v(t_2) = gt_2 - v_0 + \frac{g}{\gamma}\log\left(1 + \frac{v_0\gamma}{g}\right).$$

Show that if the viscous effect is not very large, (i.e., $\gamma \ll g/v_0$) the speed of the ball as it returns to the height from which it was first fired is given to first order in γ by

$$v(t_2) \approx v_0\left(1 - \frac{v_0\gamma}{2g}\right).$$

31. A particle with mass m and positive electric charge e is traveling in the positive direction along the negative y-axis with a constant speed v_0. At time $t = 0$, the particle reaches the point $(x, y, z) = (0, 0, 0)$ where it enters a uniform magnetic field $\mathbf{B} = \mathbf{k}\,B$ directed along the positive z-axis. The magnetic field is zero everywhere in the region corresponding to $y < 0$. Consider the only force acting on the particle to be the magnetic force

$$\mathbf{F} = e(\mathbf{v} \times \mathbf{B}).$$

Use Newton's second law in the form

$$\mathbf{F} = m\frac{d^2\mathbf{r}}{dt^2}$$

to obtain the x and y components of the equation of motion. Solve these coupled, second-order differential equations for the position coordinates $x = x(t)$ and $y = y(t)$ as explicit functions of t. Based on your results, describe completely in words the path of the particle.

32. For the particle moving in a uniform magnetic field described in Exercise 5.31, write Newton's second law in the form

$$\mathbf{F} = m\frac{d\mathbf{v}}{dt}.$$

Solve the resulting coupled, first-order differential equations for the components of the velocity $v_x(t)$ and $v_y(t)$. From these results, obtain the position coordinates $x(t)$ and $y(t)$. Compare these expressions with those obtained in Exercise 5.31.

33. In our treatment of the damped oscillator in Section 5.4.1, we considered both the case of the strong spring ($\omega_0 > \gamma$) and the case of the weak spring ($\omega_0 < \gamma$). What happens if $\omega_0 = \gamma$ exactly? In this case, the differential equation for $x(t)$ becomes

$$\frac{d^2x(t)}{dt^2} + 2\gamma \frac{dx(t)}{dt} + \gamma^2 x(t) = 0.$$

Verify that setting $x(t) = e^{ct}$ yields only one solution, namely $x(t) = Ae^{-\gamma t}$. For a second linearly independent solution, substitute $x(t) = Bte^{pt}$ into the equation and find the appropriate value for the constant p. Write $x(t)$ as a sum of these two solutions. Assume the initial conditions $x(0) = b$ and $v(0) = v_0$ to evaluate the arbitrary constants A and B in your solution. Now write expressions for position $x(t)$, velocity $v(t)$, and acceleration $a(t)$ in terms of the given parameters. Find the (finite) time t_1 at which the acceleration of the particle is zero. What are the position and velocity at this time? Calculate the forces $F_{\text{viscous}} = -2m\gamma v$ and $F_{\text{elastic}} = -m\gamma^2 x$ at time t_1. Is this result reasonable? What are the position, velocity, and acceleration of the particle after an infinitely long time? Set $v_0 = 2\gamma b$, and plot each of the quantities x/b, $v/\gamma b$, and $a/b\gamma^2$ as functions of the dimensionless quantity γt. Write a paragraph describing clearly in words the behavior of the particle throughout its motion paying particular attention to significant events (e.g., maximum displacement, changes in direction of motion, net force equal to zero, etc.). Repeat these plots with analogous descriptions of the motion for $v_0 = -2\gamma b$ and for $v_0 = 0$.

34. The equation of motion of a damped oscillator of mass m is

$$\frac{d^2x(t)}{dx^2} + 2\gamma \frac{dx(t)}{dx} + \omega_0^2 x(t) = 0.$$

Assume the particle is initially in motion at the point $x(0) = b$ with velocity $v(0) = v_0$. Using the method of solution in Section 5.4.1, we get

$$x(t) = e^{-\gamma t}\left[\frac{v_0 + b(\lambda + \gamma)}{2\lambda}e^{\lambda t} - \frac{v_0 - b(\lambda - \gamma)}{2\lambda}e^{-\lambda t}\right],$$

where we have defined a new parameter $\lambda = \sqrt{\gamma^2 - \omega_0^2}$. Obtain expressions for the velocity $v(t)$ and acceleration $a(t)$ of the particle. Take the limit $\omega_0 \to \gamma$ (i.e., $\lambda \to 0$) in this expression for $x(t)$ and compare with the corresponding result obtained in Exercise 5.33. Do the two expressions for $x(t)$ with $\omega_0 = \gamma$ agree?

35. The series method yields only one solution to the differential equation in Exercise 5.33

$$\frac{d^2x(t)}{dt^2} + 2\gamma\frac{dx(t)}{dt} + \gamma^2 x(t) = 0.$$

To obtain a general solution with the required two arbitrary constants, make the change of dependent variable $x(t) = e^{pt}g(t)$. Substitute this for $x(t)$ in the differential equation and obtain another differential equation for $g(t)$. Choose the constant p so that the $g'(t)$-term drops out of the equation. Solve the resulting differential equation for $g(t)$. Write the general solution for the original equation for $x(t)$. Compare this expression with the corresponding general solution obtained in Exercise 5.33.

36. Apply the series method of solution to the differential equation

$$x^2\frac{d^2u(x)}{dx^2} - 12x\frac{du(x)}{dx} + 36u(x) = 0$$

to obtain two linearly independent solutions of very simple form. Use these to write the general solution to this equation.

37. Show that the series method yields two very simple linearly independent solutions to the differential equation

$$x^2\frac{d^2u(x)}{dx^2} - ax\frac{du(x)}{dx} + au(x) = 0,$$

where a is any real constant. Write the general solution to this equation.

6

Partial Differential Equations

6.1   The Method of Separation of Variables

Frequently, it is *partial* differential equations that we encounter in applied mathematics. Quite often, there are symmetries or other features which make it possible to reduce the partial differential equation to two or more *ordinary* differential equations. Let us illustrate the procedure by considering an example from physics.

In the early part of the twentieth century, it became clear that Newton's physics of the seventeenth century fails to describe the behavior of very small systems of the order of the size of atoms and smaller. For such tiny objects, a new physics called quantum mechanics is required. One formulation of quantum mechanics is due to Schrödinger, in which Schrödinger's equation is the analogue to Newton's second law.[1] For a single particle of mass m moving in one dimension, this equation is

$$-\frac{\hbar^2}{2m}\frac{\partial^2}{\partial x^2}\psi(x,t) + V(x)\psi(x,t) = i\hbar\frac{\partial}{\partial t}\psi(x,t), \qquad (6.1)$$

where $\hbar$ is Planck's constant h divided by 2π i.e., $\hbar \equiv h/2\pi$. The information about the physical description of the particle is contained in the complex function, $\psi(x,t)$, called the *wave function* or *state function* for the particle and Schrödinger's equation describes the behavior of this function in space and time.

On the left-hand side of Eq. (6.1), the operators which operate on $\psi(x,t)$ depend on x (position) only, whereas the differential operator on the right-hand side

---

[1] A rationale for Schrödinger's equation will be given in Section 8.4.2. See also Appendix B.

depends only on t (time). This separation of the operators leads to solutions that are products of separate functions of x and t,

$$\psi(x, t) = u(x)f(t).$$

This can be seen by direct substitution into Eq. (6.1),

$$\frac{1}{u(x)}\left[-\frac{\hbar^2}{2m}\frac{d^2}{dx^2} + V(x)\right]u(x) = \frac{i\hbar}{f(t)}\frac{d}{dt}f(t). \qquad (6.2)$$

After substitution, we have divided Eq. (6.1) by $u(x)f(t)$ to obtain Eq. (6.2). Note that partial derivatives have been replaced by ordinary derivatives, since the differential operators now operate on functions of single variables.

We see that the left-hand side of Eq. (6.2) is a function of x only while the right-hand side is a function of t only. Since x and t are independent variables, each side of Eq. (6.2) must be equal to a constant. Let us call this constant E. From dimensional analysis, it is easily verified that E is an energy and, in fact, represents the total energy of the particle.

For the right-hand side of Eq. (6.2), we then have

$$\frac{i\hbar}{f}\frac{df}{dt} = E$$

which we can integrate immediately to obtain

$$f(t) = e^{-iEt/\hbar}.$$

Thus, the solution to Eq. (6.1) takes the form

$$\psi(x, t) = u(x)e^{-iEt/\hbar},$$

where $u(x)$ satisfies the second-order differential equation,

$$-\frac{\hbar^2}{2m}\frac{d^2}{dx^2}u(x) + (V(x) - E)u(x) = 0. \qquad (6.3)$$

This method of solving a partial differential equation is known as the *method of separation of variables*.

To summarize this method, we note that the general form of a homogeneous partial differential equation for a function $\phi(x, y)$ of two independent variables x and y is

$$\Omega_{xy}\phi(x, y) = 0, \qquad (6.4)$$

where Ω_{xy} is an operator containing partial derivatives with respect to x and y. In special cases, it may be possible to write

$$\Omega_{xy} = \Omega_x + \Omega_y,$$

where the operator Ω_x does not contain y and Ω_y does not contain x. In such cases, the differential equation can be written

$$\Omega_x\phi(x, y) = -\Omega_y\phi(x, y), \qquad (6.5)$$

and the solution can be written as a product

$$\phi(x, y) = u(x)v(y).$$

On substituting this product for $\phi(x, y)$ in Eq. (6.5) and dividing through by $u(x)v(y)$ we get

$$\frac{1}{u(x)}\Omega_x u(x) = -\frac{1}{v(y)}\Omega_y v(y). \tag{6.6}$$

The left-hand side of this equation depends only on x and the right-hand side only on y. Since x and y are independent, Eq. (6.6) can be satisfied for arbitrary x and y only if both sides are equal to the same constant. Call it λ. Then, from Eq. (6.6) we get two *ordinary* differential equations

$$\Omega_x u(x) - \lambda u(x) = 0$$

and

$$\Omega_y v(y) + \lambda v(y) = 0.$$

The constant parameter λ is called the *separation constant*. This method can be extended to functions of several variables by separating off the independent variables one variable at a time with a new separation constant introduced at each separation. Although nonlinear partial differential equations have wide application in some branches of applied mathematics,[2] we restrict our discussion in this book to linear partial differential equations.

6.2 The Quantum Harmonic Oscillator

In order to solve Eq. (6.3), one must specify the force—represented by the potential energy function $V(x)$—that governs the particle's behavior. As an example, let us consider the motion of a particle moving under the influence of a force which obeys Hooke's law,

$$F = -m\omega^2 x,$$

where we have expressed the force constant in terms of the mass m of the particle and the natural angular frequency ω of its oscillatory motion. The potential energy of the system is just the work done in displacing the particle by an amount x from its equilibrium position ($x' = 0$),

$$V(x) = -\int_0^x F(x')\,dx' = -\int_0^x -m\omega^2 x'\,dx' = \tfrac{1}{2}m\omega^2 x^2.$$

---

[2] Examples include dispersive wave phenomena, viscous fluid flow, population biology, chaos, and the propagation of solitary waves—to name just a few.

So, for the quantum harmonic oscillator, the spatial equation can be written as

$$\frac{d^2u(x)}{dx^2} + \frac{2m}{\hbar^2}\left(E - \tfrac{1}{2}m\omega^2 x^2\right)u(x) = 0. \tag{6.7}$$

Preliminary to finding a solution to this equation, we introduce a new independent variable ρ defined by

$$x \equiv b\rho.$$

The constant parameter b has the dimension of length which makes ρ dimensionless. With this change of variable, Eq. (6.7) becomes

$$\frac{d^2u(\rho)}{d\rho^2} + \left(\lambda - \frac{m^2\omega^2 b^4}{\hbar^2}\rho^2\right)u(\rho) = 0,$$

where we have defined the parameter λ by

$$\lambda \equiv \frac{2mEb^2}{\hbar^2}. \tag{6.8}$$

The length b is still at our disposal. Let us choose it to have the value

$$b \equiv \sqrt{\frac{\hbar}{m\omega}}. \tag{6.9}$$

With this choice, Eq. (6.7) takes the compact form

$$\frac{d^2u(\rho)}{d\rho^2} + \left(\lambda - \rho^2\right)u(\rho) = 0. \tag{6.10}$$

Following the procedure in Chapter 5, we look for a series solution to Eq. (6.10),

$$u(\rho) = \sum_{k=0}^{\infty} c_k \rho^{k+s}.$$

By direct substitution of this expression into Eq. (6.10), we get

$$s(s-1)c_0\rho^{s-2} + (1+s)sc_1\rho^{s-1}$$
$$+ \left[(2+s)(1+s)c_2 + \lambda c_0\right]\rho^s + \left[(3+s)(2+s)c_3 + \lambda c_1\right]\rho^{s+1}$$
$$+ \sum_{k=2}^{\infty} \left[(k+s+2)(k+s+1)c_{k+2} + \lambda c_k - c_{k-2}\right]\rho^{k+s} = 0$$

where we have combined terms with the same powers of ρ.

Notice that this result leads to a *three term* recursion formula for the coefficients

$$c_{k+2} = \frac{c_{k-2} - \lambda c_k}{(k+s+2)(k+s+1)}.$$

This is not a very useful expression, particularly if the sum is an infinite series. We need something more practical. Clearly, it is the ρ^2 term in Eq. (6.10) which causes the trouble. Let us see if we can get rid of it. We assume that $u(\rho)$ can be written as a product

$$u(\rho) = f(\rho)g(\rho)$$

and choose $f(\rho)$ so that the differential equation for $g(\rho)$ can be solved by the series method. Substituting this product into Eq. (6.10) and rearranging terms, we obtain the equation

$$\frac{d^2g}{d\rho^2} + \frac{2f'}{f}\frac{dg}{d\rho} + \left(\frac{f''}{f} + \lambda - \rho^2\right)g = 0.$$

To eliminate ρ^2 in the zero-order term, we require that $f''/f - \rho^2$ be equal to a constant. By assuming a solution of the form $f = \exp(a\rho^2)$, we find that

$$f(\rho) = e^{\pm\frac{1}{2}\rho^2}.$$

To get solutions which remain well behaved as $|\rho| \to \infty$, we choose the lower sign in the exponent. Thus, the physically acceptable solutions to Eq. (6.10) have the form

$$u(\rho) = e^{-\frac{1}{2}\rho^2}g(\rho), \tag{6.11}$$

where $g(\rho)$ satisfies the differential equation,

$$\frac{d^2g}{d\rho^2} - 2\rho\frac{dg}{d\rho} + (\lambda - 1)g = 0. \tag{6.12}$$

Again, we look for a series solution,

$$g(\rho) = \sum_{k=0}^{\infty} a_k\rho^{k+t}. \tag{6.13}$$

Substituting this series into Eq. (6.12) and collecting terms, we obtain

$$t(t-1)a_0\rho^{t-2} + (1+t)ta_1\rho^{t-1}$$
$$+ \sum_{k=0}^{\infty}\left[(k+t+2)(k+t+1)a_{k+2} - \left[2(k+t) - \lambda + 1\right]a_k\right]\rho^{k+t} = 0,$$

which holds for arbitrary values of ρ. Since terms in different powers of ρ are linearly independent, we have

$$t(t-1)a_0 = 0 \qquad\qquad\qquad (1+t)ta_1 = 0$$

and

$$(k+t+2)(k+t+1)a_{k+2} - \left[2(k+t) - \lambda + 1\right]a_k = 0. \tag{6.14}$$

We choose t such that the first two equations here are satisfied and either a_0 or a_1 or both are arbitrary. Then we use the two-term recursion formula in Eq. (6.14) to generate the rest of the coefficients a_k for $k \geq 2$ to determine the solution $g(\rho)$, and hence $u(\rho)$, completely. The physical boundary conditions will determine the values of a_0 and a_1.

Let us examine these solutions in more detail. For definiteness, we take $t = 1$ in Eq. (6.14). In this case, we must have $a_1 = 0$ and $a_0 \neq 0$. Examining Eq. (6.13), we note that for $|\rho| \gg 1$, it is the terms for larger values of k that dominate the infinite series in Eq. (6.13). For large values of k, the recursion formula Eq. (6.14) becomes

$$\lim_{k \to \infty} a_{k+2} \to \frac{2}{k+2} a_k.$$

By successive applications of this formula, we find that for large k,

$$a_k \approx \frac{1}{(k/2)!} a_0.$$

Substituting this approximation into Eq. (6.13), we get

$$g(\rho) = \sum_{\substack{k=0 \\ k=even}}^{\infty} a_k \rho^{k+1} = a_0 \sum_{\substack{k=0 \\ k=even\equiv 2n}}^{\infty} \frac{\rho^{k+1}}{(k/2)!} = a_0 \sum_{n=0}^{\infty} \frac{\rho^{2n+1}}{n!} = a_0 \rho e^{\rho^2}.$$

Thus, for large values of $|\rho|$ we have

$$g(\rho) \underset{|\rho| \to \infty}{\to} a_0 \rho e^{\rho^2}.$$

Inserting this result into Eq. (6.11), we obtain for the asymptotic behavior of the solution to Eq. (6.10),

$$u(\rho) \underset{|\rho| \to \infty}{\to} a_0 \rho e^{\frac{1}{2}\rho^2} \to \pm\infty.$$

Since $\rho = x/b$ measures the position x of the particle relative to its equilibrium position at $x = 0$, this is not acceptable behavior for the solution to the time-independent Schrödinger equation, Eq. (6.3). So we impose the *physical* requirement that the sum on the right-hand side of Eq. (6.13) cannot be an infinite series. It must terminate for some finite value of k. The sum is a polynomial in ρ. In the case we have chosen [$t = 1$ in Eq. (6.14)], the polynomial Eq. (6.13) contains only odd powers of ρ.

For the series in Eq. (6.13) to be a polynomial means that for some value of k, say $k = k_1$, we must have $a_{k_1+2} = 0$ with $a_{k_1} \neq 0$. With this requirement, we see that Eq. (6.14) implies

$$2(k_1 + 1) - \lambda + 1 = 2k_1 + 3 - \lambda = 0. \tag{6.15}$$

From the definitions of λ and b given in Eqs. (6.8) and (6.9), respectively, we find that λ is related to the total energy E of the particle by $\lambda = 2E/\hbar\omega$. Substituting this result for λ in Eq. (6.15), we see that the total energy E is discrete or *quantized* according to

$$E = (k_1 + \tfrac{3}{2})\hbar\omega \quad \text{with} \quad k_1 = 0, 2, 4, \ldots \tag{6.16}$$

and the series in Eq. (6.13) is a polynomial in odd powers of ρ.

On repeating this analysis for $t = -1$, it becomes clear that Eq. (6.14) requires for some value of k, call it k_2, that $a_{k_2+2} = 0$, but $a_{k_2} \neq 0$, again leading to discrete values for the total energy E,

$$E = (k_2 - \tfrac{1}{2})\hbar\omega \quad \text{with} \quad k_2 = 1, 3, 5, \ldots. \tag{6.17}$$

In this case, the series in Eq. (6.13) is a polynomial in even powers of ρ.

With the definitions $n \equiv k_1 + 1$ in Eq. (6.16) and $n \equiv k_2 - 1$ in Eq. (6.17), these equations can be combined into a single expression for the energy spectrum,

$$E_n = (n + \tfrac{1}{2})\hbar\omega \qquad \text{with} \quad n = 0, 1, 2, 3, \ldots. \tag{6.18}$$

The corresponding solution to Eq. (6.7) has the form

$$u_n(x) = \{n\text{th-degree polynomial in } x\} \cdot e^{-x^2/2b^2}. \tag{6.19}$$

If we take $t = 0$ in Eq. (6.14), in general, both a_0 and a_1 can be arbitrary. However, if we require physically acceptable solutions, the series in Eq. (6.13) must be a polynomial. We have seen that if we arrange for $a_0 \neq 0$ so that Eq. (6.14) generates a finite number of even-indexed coefficients a_k up to $k = n$ with all the even-indexed coefficients for $k > n$ equal to zero we obtain a polynomial of degree n. However, starting with a_1, the recursion formula Eq. (6.14) will yield an infinite number of odd-indexed coefficients. That is, the corresponding series will diverge for large values of $|\rho|$ and is therefore physically unacceptable. Thus, in this case we must set $a_1 = 0$. A similar argument follows for $t = 0$ with $a_1 \neq 0$ and $n =$ an odd integer. In this case, the series in Eq. (6.13) is a polynomial in odd powers of ρ and physically acceptable solutions to Eq. (6.7) require that $a_0 = 0$.

6.3 A Conducting Sphere in an Electric Field

Let us now look at another simple case in which a partial differential equation arises. We recall from Chapter 3 the notion of flux. According to the divergence theorem,

$$\oint_S \mathbf{F} \cdot d\mathbf{A} = \int_V \nabla \cdot \mathbf{F}\, d^3r,$$

where the left-hand side represents the flux of the vector field $\mathbf{F}$ out over the closed surface S which is the boundary to the volume V. For the special case where $\nabla \cdot \mathbf{F} = 0$, we see that the net flux out over the surface S of *any* volume V is zero. This means that all flux lines representing the field $\mathbf{F}$ are closed loops. There are no *sources* of the field $\mathbf{F}$ if $\nabla \cdot \mathbf{F} = 0$. This is the case for the magnetic field. There are no magnetic monopoles, hence for the magnetic flux density $\mathbf{B}$ we have $\nabla \cdot \mathbf{B} = 0$.

Now, electric monopoles (charges) *do* exist so that, in general, for the electric flux density $\mathbf{E}$ we have $\nabla \cdot \mathbf{E} \neq 0$.[3] But if we consider a region that does not contain any electric charge, then at each point in the region

$$\nabla \cdot \mathbf{E} = 0.$$

In the electrostatic case, an electric field is produced by a distribution of static electric charges. It is a linear superposition of fields due to the individual charges.

---

[3]See Eq. (3.20).

The electric field $\mathbf{E}$ due to a single-point charge q is obtained from Coulomb's law. We have

$$\mathbf{E} = \frac{kq}{r^3}\mathbf{r},$$

where $\mathbf{r}$ is the vector which locates the position of the field point with respect to the charge. If we treat $\mathbf{r}/r^3$ as a product of $\mathbf{r}$ and r^{-3} and use rectangular coordinates, we find by explicit differentiation that

$$\nabla \times \mathbf{E} = 0.[4]$$

Now, the curl of the gradient of any scalar is identically zero. That is, for any scalar function $U(\mathbf{r})$,

$$\nabla \times \nabla U \equiv 0.$$

Therefore, we can define any electrostatic field $\mathbf{E}$ by

$$\mathbf{E} = -\nabla U$$

and $\nabla \times \mathbf{E} = 0$ is automatically satisfied. On requiring that $\nabla \cdot \mathbf{E} = 0$, we see that the electric potential U must satisfy the partial differential equation

$$\nabla^2 U = 0. \tag{6.20}$$

This equation is known as *Laplace's equation*.

6.3.1   BOUNDARY CONDITIONS

As a specific application, we consider the problem posed by placing an uncharged, conducting sphere of radius R in a uniform electric field $\mathbf{E}_0$. In the static case, there is no electric current. We choose the origin of our coordinate system at the center of the sphere. Since the charges are free to move in the conductor, components of the electric field tangent to the surface of the sphere must vanish there. Otherwise, currents arise from the electric forces on the charges in the conductor. So, the charges in the sphere are redistributed by the external field in such a way that $\mathbf{E}$ is perpendicular to the surface of the sphere at $r = R$. In other words,

$$E_{\text{tangential}}\Big|_{r=R} = 0. \tag{6.21}$$

We expect that the field is distorted by the sphere only in the neighborhood around the sphere. Therefore, at large distances from the sphere, the electric field should approach $\mathbf{E}_0$. If we choose the z-axis along the direction of the external field as indicated in Fig. 6.1, we can express this condition as

$$\mathbf{E} \underset{r\to\infty}{\to} kE_0. \tag{6.22}$$

These latter two relations are called *the boundary conditions*.

Our problem consists of three parts:

---

[4]See Exercise 6.12.

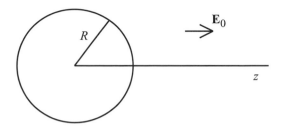

FIGURE 6.1. An uncharged conducting sphere in a uniform electric field.

- Solve Laplace's equation for $U(\mathbf{r})$.
- Invoke the boundary conditions.
- Calculate the electric field from $\mathbf{E} = -\nabla U$.

6.3.2 SEPARATION OF VARIABLES

Because of the axial symmetry evident here, it is reasonable to look for solutions to Eq. (6.20) in spherical polar coordinates. We want to express $U(\mathbf{r})$ as an explicit function of (r, θ, ϕ). In this case, Eq. (6.20) is written as[5]

$$\frac{1}{r^2}\frac{\partial}{\partial r}\left(r^2\frac{\partial U}{\partial r}\right) + \frac{1}{r^2 \sin\theta}\frac{\partial}{\partial\theta}\left(\sin\theta\frac{\partial U}{\partial\theta}\right) + \frac{1}{r^2 \sin^2\theta}\frac{\partial^2 U}{\partial\phi^2} = 0. \qquad (6.23)$$

In the problem at hand, we have symmetry about the z-axis which means that U has no ϕ-dependence. That is, $U = U(r, \theta)$ and Eq. (6.20) reduces to

$$\frac{\partial}{\partial r}\left(r^2\frac{\partial U}{\partial r}\right) = -\frac{1}{\sin\theta}\frac{\partial}{\partial\theta}\left(\sin\theta\frac{\partial U}{\partial\theta}\right). \qquad (6.24)$$

In this form, it is clear that the equation can be separated into r-dependent and θ-dependent parts by writing

$$U(r, \theta) = R(r)S(\theta).$$

With this substitution into Eq. (6.24), we obtain

$$\frac{1}{R(r)}\frac{d}{dr}\left(r^2\frac{dR(r)}{dr}\right) = -\frac{1}{S(\theta)\sin\theta}\frac{d}{d\theta}\left(\sin\theta\frac{dS(\theta)}{d\theta}\right).$$

This equation must hold for arbitrary values of r and θ. Since r and θ are independent, the two sides must be separately equal to the same constant, which we call λ. This leads to the two ordinary differential equations,

$$\frac{d^2 S}{d\theta^2} + \cot\theta\frac{dS}{d\theta} + \lambda S = 0, \qquad (6.25)$$

and

$$\frac{d}{dr}\left(r^2\frac{dR}{dr}\right) - \lambda R = 0. \qquad (6.26)$$

---

[5] See Appendix A.3.2.

Let us begin with Eq. (6.25). For the series method of solution to be successful we need rational coefficients in each term. Otherwise, we shall not obtain a useful recursion formula for the coefficients in the series expansion. Clearly, $\cot\theta$ does not meet this requirement. If we change the independent variable to $x = \cos\theta$, then with a slight change in notation defining $S(\theta) \equiv P(\cos\theta)$, Eq. (6.25) can be written as

$$(1 - x^2)\frac{d^2 P}{dx^2} - 2x\frac{dP}{dx} + \lambda P = 0. \tag{6.27}$$

This is known as *Legendre's equation.*

To obtain a series solution, we substitute

$$P(x) = \sum_{k=0}^{\infty} a_k x^{k+s} \tag{6.28}$$

into Eq. (6.27). Collecting terms in like powers of x, we get

$$s(s - 1)a_0 x^{s-2} + (s + 1)sa_1 x^{s-1}$$
$$+ \sum_{k=0}^{\infty} \left[(k + s + 2)(k + s + 1)a_{k+2} - [(k + s)(k + s + 1) - \lambda]a_k\right]x^{k+s} = 0.$$

The terms corresponding to different powers of x are linearly independent. Therefore, coefficients of the different powers of x must vanish separately. We get the two *indicial* equations

$$s(s - 1)a_0 = 0 \qquad \text{and} \qquad (s + 1)sa_1 = 0$$

and the recursion formula

$$(k + s + 2)(k + s + 1)a_{k+2} - \left((k + s)(k + s + 1) - \lambda\right)a_k = 0.$$

From this latter equation, it is clear that

$$\frac{a_{k+2}}{a_k} = \left[\frac{(k + s)(k + s + 1) - \lambda}{(k + s + 2)(k + s + 1)}\right]_{k\to\infty} \longrightarrow 1$$

which tells us that the series does *not* converge for $x = 1$. Now, $x = 1$ corresponds to a point on the polar axis which is a physically allowed point in space ($\theta = 0$). What do we do? To obtain a finite solution, the series in Eq. (6.28) must terminate. It must be a polynomial. This means that for some k, say $k = n$, $a_n \neq 0$, but $a_{n+2} = 0$. This requires that we have

$$\lambda = (n + s)(n + s + 1).$$

Now, suppose n is an even integer. In this case $a_{n+2} = 0$, but we still have an infinite series of terms with the coefficients of the powers of x having *odd* indices. For nontrivial solutions then, $a_1 = 0$ and $a_0 \neq 0$. Similarly, if n is an odd integer, then $a_0 = 0$ and $a_1 \neq 0$.

Let us consider the case for n being an even integer. If we choose $s = 0$, $a_0 \neq 0$, and $a_1 = 0$, then the indicial equations are satisfied and

$$\lambda = n(n + 1). \tag{6.29}$$

In this case, it is clear that the series in Eq. (6.28) is a polynomial containing only even powers of x. That is,

$$P_n(x) = \sum_{j=0}^{\frac{n}{2}} c_{2j} x^{2j}.$$

Similarly, for n an odd integer, we choose $s = 0$, $a_1 \neq 0$, and $a_0 = 0$. Again $\lambda = n(n + 1)$. This time we obtain a polynomial in *odd* powers of x,

$$P_n(x) = \sum_{j=0}^{\frac{n-1}{2}} c_{2j+1} x^{2j+1}.$$

With a certain normalization, the polynomials $P_n(x)$ are called *Legendre polynomials*. The first few of these are given below.

Legendre Polynomials

$$P_0(x) = 1$$
$$P_1(x) = x$$
$$P_2(x) = \tfrac{1}{2}(3x^2 - 1)$$
$$P_3(x) = \tfrac{1}{2}(5x^3 - 3x)$$
$$P_4(x) = \tfrac{1}{8}(35x^4 - 30x^2 + 3)$$
$$P_5(x) = \tfrac{1}{8}(63x^5 - 70x^3 + 15x)$$

With the constraint on λ given in Eq. (6.29), the equation for $R(r)$, Eq. (6.26), becomes

$$\frac{d}{dr}\left(r \frac{dR(r)}{dr}\right) - n(n + 1)R(r) = 0. \tag{6.30}$$

The structure of this equation suggests that we look for solutions of the form $R(r) = r^p$, where p is a constant to be determined. Substituting this expression into Eq. (6.30) yields

$$\left[p(p + 1) - n(n + 1)\right]r^p = 0 \qquad \text{for all } r.$$

Setting the factor in brackets equal to zero, we get two solutions for p, $p = n$ and $p = -n - 1$. Thus, the general solution to Eq. (6.30) for a given value of n can be written

$$R(r) = A_n r^n + B_n r^{-n-1}.$$

Finally, the general solution to Eq. (6.24) takes the form

$$U(r, \theta) = \sum_{n=0}^{\infty} \left(A_n r^n + B_n r^{-n-1}\right) P_n(\cos \theta). \tag{6.31}$$

6.3.3   IMPOSING THE BOUNDARY CONDITIONS

Now let us translate the physical boundary conditions on the field expressed in Eqs. (6.21) and (6.22) into constraints on the potential $U(r, \theta)$. For Eq. (6.21), we write

$$E_{\text{tangential}}\Big|_{r=R} = -(\nabla U)_\theta\Big|_{r=R} = -\left[\frac{1}{r}\frac{\partial U}{\partial \theta}\right]_{r=R} = 0,$$

which implies that

$$U(R, \theta) \text{ is independent of } \theta. \tag{6.32}$$

In Eq. (6.22), we have

$$\mathbf{E} = -\nabla U = -\mathbf{i}(\nabla U)_x - \mathbf{j}(\nabla U)_y - \mathbf{k}(\nabla U)_z \underset{r\to\infty}{\longrightarrow} \mathbf{E}_0 = \mathbf{k}E_0.$$

Comparing coefficients of the unit vector $\mathbf{k}$, we see that

$$-(\nabla U)_z = -\frac{\partial U}{\partial z} \underset{r\to\infty}{\longrightarrow} E_0.$$

We can integrate this latter equation directly to obtain

$$U(r, \theta) \underset{r\to\infty}{\longrightarrow} -E_0 z + U_0 = -E_0 r \cos\theta + U_0, \tag{6.33}$$

where U_0 is an arbitrary constant of integration.

Next, we apply the condition expressed in Eq. (6.33) to $U(r, \theta)$. As a first step, we rewrite Eq. (6.31) by setting apart the first two terms in the sum,

$$\begin{aligned} u(r, \theta) &= A_0 + \frac{B_0}{r} + \left(A_1 r + \frac{B_1}{r^2}\right)\cos\theta + \sum_{n=2}^{\infty}\left(A_n r^n + \frac{B_n}{r^{n+1}}\right)P_n(\cos\theta) \\ &\underset{r\to\infty}{\longrightarrow} A_0 + A_1 r \cos\theta + \sum_{n=2}^{\infty} A_n r^n P_n(\cos\theta) \\ &= -E_0 r \cos\theta + U_0, \end{aligned}$$

where we have taken the Legendre polynomials $P_0(\cos\theta)$ and $P_1(\cos\theta)$ from the list on page 113. Rewriting this last equation, we have

$$A_0 - U_0 + (A_1 + E_0)r \cos\theta + \sum_{n=2}^{\infty} A_n r^n P_n(\cos\theta) = 0. \tag{6.34}$$

Because the Legendre polynomials are linearly independent functions of $\cos\theta$, the coefficient of each Legendre polynomial in Eq. (6.34) must be set equal to zero. From this condition, we get

$$\begin{aligned} A_0 &= U_0, \\ A_1 &= -E_0, \\ A_n &= 0 \quad \text{for } n \geq 2. \end{aligned}$$

With these results, our expression for $U(r, \theta)$ in Eq. (6.31) now reads

$$U(r, \theta) = U_0 + \frac{B_0}{r} + \left(-E_0 r + \frac{B_1}{r^2}\right)\cos\theta + \sum_{n=2}^{\infty} \frac{B_n}{r^{n+1}} P_n(\cos\theta). \qquad (6.35)$$

According to Eq. (6.32), the right-hand side of Eq. (6.35) must be independent of θ when $r = R$. Again invoking the linear independence of the Legendre polynomials $P_n(\cos\theta)$, we must require that

$$B_1 = E_0 R^3$$
$$B_n = 0 \quad \text{for } n \geq 2.$$

The potential $U(r, \theta)$, Eq. (6.35), now takes the form

$$U(r, \theta) = U_0 + \frac{B_0}{r} + \left(\frac{R^3}{r^2} - r\right)E_0 \cos\theta.$$

To evaluate the remaining constant B_0, we again resort to the physics and recognize that the term B_0/r in the potential corresponds to a contribution to the electric field arising from an electric charge residing on the sphere.[6] Since the sphere is uncharged in the example we are considering, we set $B_0 = 0$. Our final expression for the electric potential for points outside the sphere is

$$U(r, \theta) = U_0 + \left(\frac{R^3}{r^2} - r\right)E_0 \cos\theta. \qquad (6.36)$$

From this potential, we obtain the electric field for all points outside the uncharged sphere,

$$\mathbf{E} = -\nabla U = -\hat{\mathbf{r}}\left(\frac{\partial U(r, \theta)}{\partial r}\right) - \hat{\boldsymbol{\theta}}\left(\frac{1}{r}\frac{\partial U(r, \theta)}{\partial \theta}\right)$$
$$= \hat{\mathbf{r}}\left(\frac{2R^3}{r^3} + 1\right)E_0 \cos\theta + \hat{\boldsymbol{\theta}}\left(\frac{R^3}{r^3} - 1\right)E_0 \sin\theta. \qquad (6.37)$$

All arbitrary constants have been fixed by the boundary conditions. The field depends only on the size of the sphere, the strength of the external electric field, and the position of the field point.

6.4   The Schrödinger Equation for a Central Field

For a small particle of mass m, the time-independent Schrödinger equation in three dimensions is[7]

$$\left[\frac{-\hbar^2}{2m}\nabla^2 + V(\mathbf{r})\right]u(\mathbf{r}) = Eu(\mathbf{r}). \qquad (6.38)$$

---

[6]In this case, set $B_0 = kQ$ where Q is the total charge on the sphere.
[7]See Appendix B.4.

If the potential energy of the particle depends only on the distance r from the force center, i.e., $V(\mathbf{r}) = V(r)$, it is convenient to write the operator ∇^2 in spherical polar coordinates.[8] Hence, Eq. (6.38) becomes

$$\frac{-\hbar^2}{2m}\left[\frac{1}{r^2}\frac{\partial}{\partial r}\left(r^2\frac{\partial}{\partial r}\right) + \frac{1}{r^2\sin\theta}\frac{\partial}{\partial\theta}\left(\sin\theta\frac{\partial}{\partial\theta}\right) + \frac{1}{r^2\sin^2\theta}\frac{\partial^2}{\partial\phi^2}\right]u(\mathbf{r})$$
$$+ V(r)u(\mathbf{r}) = Eu(\mathbf{r}).$$

After some rearrangement, we get,

$$\left[\frac{\partial}{\partial r}\left(r^2\frac{\partial}{\partial r}\right) + \frac{2mr^2}{\hbar^2}(E - V(r))\right]u(\mathbf{r})$$
$$= -\left[\frac{1}{\sin\theta}\frac{\partial}{\partial\theta}\left(\sin\theta\frac{\partial}{\partial\theta}\right) + \frac{1}{\sin^2\theta}\frac{\partial^2}{\partial\phi^2}\right]u(\mathbf{r}). \qquad (6.39)$$

Clearly, the operator on the left-hand side of this equation operates only with respect to r on $u(\mathbf{r}) = u(r, \theta, \phi)$ and on the right-hand side the operator depends only on the angular variables θ and ϕ. With this separation of the operators, we can write

$$u(\mathbf{r}) = R(r)Y(\theta, \phi). \qquad (6.40)$$

Substituting this expression into Eq. (6.39) yields

$$\frac{1}{R(r)}\left[\frac{\partial}{\partial r}\left(r^2\frac{\partial}{\partial r}\right) + \frac{2mr^2}{\hbar^2}(E - V(r))\right]R(r)$$
$$= -\frac{1}{Y(\theta, \phi)}\left[\frac{1}{\sin\theta}\frac{\partial}{\partial\theta}\left(\sin\theta\frac{\partial}{\partial\theta}\right) + \frac{1}{\sin^2\theta}\frac{\partial^2}{\partial\phi^2}\right]Y(\theta, \phi). \qquad (6.41)$$

The left-hand side of Eq. (6.41) depends only on r and the right hand side only on the angles θ and ϕ. Because all three variables are independent, each side must be equal to the same constant, which we denote by λ. Equation (6.41) leads to an ordinary differential equation

$$\frac{d}{dr}\left(r^2\frac{dR(r)}{dr}\right) + \left[\frac{2mr^2}{\hbar^2}(E - V(r)) - \lambda\right]R(r) = 0. \qquad (6.42)$$

and a partial differential equation

$$\left[\frac{1}{\sin\theta}\frac{\partial}{\partial\theta}\left(\sin\theta\frac{\partial}{\partial\theta}\right) + \frac{1}{\sin^2\theta}\frac{\partial^2}{\partial\phi^2}\right]Y(\theta, \phi) + \lambda Y(\theta, \phi) = 0. \qquad (6.43)$$

Following an argument similar to the one leading to Eq. (6.29), we find that for physically allowed values of $u(\mathbf{r})$, the constant λ in Eq. (6.43) must be equal to a product of two successive integers. That is,

$$\lambda = l(l + 1), \qquad l = 0, 1, 2\ldots. \qquad (6.44)$$

---

[8] See Appendix A.3.2

Finally, for the radial equation Eq. (6.42), we get

$$\frac{1}{r^2}\left[\frac{d}{dr}\left(r^2\frac{dR_l(r)}{dr}\right)\right] + \left[\frac{2m}{\hbar^2}(E - V(r)) - \frac{l(l+1)}{r^2}\right]R_l(r) = 0, \qquad (6.45)$$

where we have written the function $R_l(r)$ with a subscript to indicate that the solutions are different for different values of the parameter l.

It is often convenient to write this equation in a slightly different form. This is achieved by defining a new function, $g_l(r) = r R_l(r)$, that results in a transformation of Eq. (6.45) to

$$r^2\frac{d^2 g_l(r)}{dr^2} + \left[\frac{2mr^2}{\hbar^2}(E - V(r)) - l(l+1)\right]g_l(r) = 0. \qquad (6.46)$$

The Schrödinger equation for a free particle, i.e., for $V(r) = 0$ is of particular interest. This will provide us with a useful example in which we can use the series method to obtain solutions to the radial equation in terms of well-known functions. First, we set $V(r) = 0$ in Eq. (6.46) and replace r by a dimensionless independent variable x according to $x = kr$, where the parameter k is related to the total energy E by $k \equiv \sqrt{2mE/\hbar^2}$. For the free particle, Eq. (6.46) now takes the form

$$x^2\frac{d^2 g_l(x)}{dx^2} + \left[x^2 - l(l+1)\right]g_l(x) = 0. \qquad (6.47)$$

Following the procedure described in Chapter 5 for solving a differential equation by the series method, we make the substitution

$$g_l(x) = \sum_{m=0}^{\infty} a_m x^{m+s}$$

in Eq. (6.47) to obtain the relations

$$\left[s(s-1) - l(l+1)\right]a_0 = 0$$
$$\left[(1+s)s - l(l+1)\right]a_1 = 0$$
$$\left[(m+s+2)(m+s+1) - l(l+1)\right]a_{m+2} + a_m = 0. \qquad (6.48)$$

From the first of these equations, we see that if $a_0 \neq 0$, either $s = -l$ or $s = l+1$. Similarly, for $a_1 \neq 0$ the second equation requires that either $s = l$ or $s = -l-1$.

With $s = l+1$, we obtain a solution that is regular at $r = 0$. This choice also requires that $a_1 = 0$. Hence, for a nontrivial solution, $a_0 \neq 0$. According to Eq. (6.48), the coefficients a_m are then given by

$$a_m = \frac{(-1)^{\frac{m}{2}}}{2(4)\cdots(m-2)m(2l+3)(2l+5)\cdots(m+2l+1)}a_0,$$
$$= \frac{(-1)^{\frac{m}{2}}(2l+1)!!}{2^{\frac{m}{2}}(\frac{m}{2})!(m+2l+1)!!}a_0, \qquad \text{(for } m \text{ even)}. \qquad (6.49)$$

Of course, from Eq. (6.48) it follows that $a_m = 0$ for m odd, since $a_1 = 0$. In arriving at the last expression in Eq. (6.49), we have used the definition

$(2p + 1)!! \equiv (1)(3)(5) \cdots (2p + 1)$. Changing the summation index to $n = m/2$, the corresponding solution to Eq. (6.47) is

$$g_l(x) = a_0(2l + 1)!! \sum_{n=0}^{\infty} \frac{(-1)^n}{2^n n!(2n + 2l + 1)!!} x^{2n+l+1}. \tag{6.50}$$

It is clear from Eq. (6.50) that $g_l(0) = 0$. Given the definition $r R_l(r) = g_l(r)$, this behavior of $g_l(r)$ assures that the corresponding solution $R_l(r)$ to the radial equation Eq. (6.45) is regular at $r = 0$.

The series in Eq. (6.50) is related to a *Bessel function* $J_\nu(x)$ of half-odd-integer order,

$$J_{l+\frac{1}{2}}(x) = \sqrt{\frac{2}{\pi}} \sum_{n=0}^{\infty} \frac{(-1)^n}{2^n n!(2n + 2l + 1)!!} x^{2n+l+\frac{1}{2}}. \tag{6.51}$$

Thus, the solution to Eq. (6.47) regular at $x = 0$ is

$$g_l(x) = a_0(2l + 1)!! \sqrt{\frac{\pi x}{2}} J_{l+\frac{1}{2}}(x) \equiv a_0(2l + 1)!! \, x \, j_l(x), \tag{6.52}$$

where the function $j_l(x)$ defined by

$$j_l(x) = \sqrt{\frac{\pi}{2x}} J_{l+\frac{1}{2}}(x) \tag{6.53}$$

is called a *spherical Bessel function* of order l.

Finally, with the definition $x = kr$, the solution to the radial equation Eq. (6.45) regular at $r = 0$ can be written

$$R_l(r) = b_l \, j_l(kr),$$

where b_l is an arbitrary constant independent of r.

It is left as an exercise for the reader to show that a solution to Eq. (6.45) irregular at $r = 0$ is

$$R_l(r) = c_l \, n_l(kr) \tag{6.54}$$

where c_l is a constant and the *spherical Neumann function* $n_l(x)$ is defined in terms of a Bessel function according to

$$n_l(x) = \sqrt{\frac{\pi}{2x}} J_{-l-\frac{1}{2}}(x), \tag{6.55}$$

with the Bessel function given by

$$J_{-l-\frac{1}{2}}(x) = \sqrt{\frac{2}{\pi}} \sum_{n=0}^{\infty} \frac{(-1)^n}{2^n n!(2n - 2l - 1)!!} x^{2n-l-\frac{1}{2}}. \tag{6.56}$$

Thus, the general solution to the radial equation for a free particle, Eq. (6.45), is

$$R_l(r) = b_l \, j_l(kr) + c_l \, n_l(kr). \tag{6.57}$$

6.5   Exercises

1. The total relativistic energy E of a particle of rest mass m and momentum $\mathbf{p}$ is given by

$$E^2 = p^2 c^2 + m^2 c^4.$$

Convert this equation into a partial differential equation for a complex function $\psi(\mathbf{r}, t)$ with substitutions of differential operators according to

$$E \rightarrow i\hbar \frac{\partial}{\partial t} \qquad \text{and} \qquad \mathbf{p} \rightarrow -i\hbar\nabla.$$

Use the method of separation of variables to reduce this equation to a partial differential equation in the space coordinates and an ordinary differential equation in time. Define clearly any functions, variables, or constants you introduce. Obtain solutions to the ordinary differential equation in t.

2. For an elastically bound particle of mass m and with potential energy $\frac{1}{2}m\omega^2 x^2$, we found that solutions to Schrödinger's equation take the form

$$\psi(x, t) = e^{-iEt/\hbar} e^{-\frac{1}{2}\rho^2} g(\rho)$$

where $g(\rho)$ satisfies the differential equation

$$\frac{d^2 g}{d\rho^2} - 2\rho\frac{dg}{d\rho} + (\lambda - 1)g = 0.$$

Here we have defined the following quantities

$$\rho \equiv \sqrt{\frac{m\omega}{\hbar}}x \qquad \text{and} \qquad \lambda \equiv \frac{2E}{\hbar\omega}.$$

For this differential equation, assume a solution of the form

$$g(\rho) = \sum_{k=0}^{\infty} a_k \rho^{k+s}.$$

Choose s so that your solution contains *only* odd powers of ρ. Require that this solution be a polynomial. What constraint does this impose on the energy E of the particle? Explain.

3. Consider the partial differential equation

$$b^2 \frac{\partial^2 \phi(x, t)}{\partial x^2} - 2x\frac{\partial \phi(x, t)}{\partial x} - \frac{b}{c}\frac{\partial \phi(x, t)}{\partial t} = 0,$$

where x and t are real, nonnegative variables and b and c are real, positive constants. Use the method of separation of variables to obtain solutions of the form

$$\phi(x, t) = f(x)g(t).$$

Choose the separation constant so that $g(t)$ decreases monotonically to zero as t approaches infinity. Write down the constraints this imposes on the separation constant. Use the series method to solve the differential equation for $f(x)$. State clearly the conditions under which $f(x)$ is a polynomial. Write down an expression for the separation constant that satisfies all of the constraints now imposed on it. Be very clear. Write down $\phi(x, t)$ for the case where $f(x)$ is a fifth-degree polynomial.

4. The time-dependent Schrödinger equation for a free particle moving in one dimension is

$$\frac{-\hbar^2}{2m} \frac{\partial^2}{\partial x^2} \psi(x, t) = i\hbar \frac{\partial}{\partial t} \psi(x, t).$$

Use the method of separation of variables to obtain two *ordinary* differential equations in position x and time t. What is the physical interpretation of the separation constant? Obtain solutions to each of these differential equations. Express your answers in terms of the given parameters and the separation constant. From your results, write down an expression for $\psi(x, t)$.

5. The function $\phi(x, t)$ is a real function of the real variables x (position) and t (time). The behavior of the function for $-L \leq x \leq L$ and $t \geq 0$ is described by the partial differential equation

$$Lx \frac{\partial^2 \phi(x, t)}{\partial x^2} + (L - x) \frac{\partial \phi(x, t)}{\partial x} - \tau \frac{\partial \phi(x, t)}{\partial t} = 0,$$

where L and τ are real, positive constant parameters. Use the method of separation of variables to obtain a solution of the form

$$\phi(x, t) = f(t)u(x).$$

Choose the separation constant so that $\phi(x, t)$ is a monotonically decreasing function of time. Use the series method to solve the differential equation for $u(x)$. State clearly and completely the conditions under which $u(x)$ is a polynomial of degree n. At time $t = 0$, the function $\phi(x, t)$ has the value λ at the point $x = 0$. Write down $\phi(x, t)$ explicitly for the case where $\phi(x, 0)$ is a fourth-degree polynomial in x. Express your result in terms of the given parameters. What is the physical interpretation of the parameter τ? For this case ($n = 4$), obtain an explicit expression for $\phi(L, t)$ in terms of the given parameters. Obtain a similar expression for $\phi(-L, t)$.

6. The time-independent Schrödinger equation for a free particle moving in three dimensions is

$$\frac{-\hbar^2}{2m} \nabla^2 u(\mathbf{r}) = E u(\mathbf{r}),$$

where E is the total energy of the particle. Use the method of separation of variables to obtain three *ordinary* differential equations in the polar coordinates r, θ, and ϕ, respectively. What are the constraints on the separation constants? Explain clearly why these constraints are required in each case.

7. The partial differential equation

$$\frac{\partial^2 T(x, t)}{\partial x^2} = \mu \frac{\partial T(x, t)}{\partial t},$$

is known as the one-dimensional *diffusion equation*. The function $T(x, t)$ may represent, for example, the concentration of a chemical diffusing through a medium. The equation may also describe the behavior of the temperature T in matter where no heat reservoirs exist. In that case, it is called the *heat flow equation*. The variables x and t are the usual position and time coordinates. Consider the parameter μ to be constant and use the method of separation of variables to obtain solutions to this differential equation. Choose the separation constant such that the solutions monotonically approach zero as time increases toward infinity. Impose the boundary conditions $T = 0$ for $x = 0$ and for $x = b$, where b is a real, positive constant. Write down explicitly the solutions (except for an overall multiplicative constant) that satisfy these boundary conditions. Explain your reasoning clearly and completely.

8. The function $u(x, t)$ is a real function of the real variables x (position) and t (time). The behavior of the function for $-b \le x \le b$ is described by the partial differential equation

$$(b^2 - x^2)\frac{\partial^2 u(x, t)}{\partial x^2} - x\frac{\partial u(x, t)}{\partial x} - \frac{1}{\omega^2}\frac{\partial^2 u(x, t)}{\partial t^2} = 0,$$

where ω is a real, positive constant parameter. Show clearly that the method of separation of variables can be used to obtain a solution of the form

$$u(x, t) = P(x)Q(t).$$

Choose the separation constant so that $u(x, t)$ is an oscillating function of time. Require that $u(x, 0) = u(x, \frac{\pi}{\omega}) = 0$. For the equation for $P(x)$, assume a series solution,

$$P(x) = \sum_{k=0}^{\infty} a_k x^{k+s}.$$

Require that $P(x)$ be a fifth-degree polynomial in x. Impose the boundary condition $u(b, \frac{\pi}{2\omega}) = b$. Using all of the constraints imposed on it, write down the complete expression for $u(x, t)$. Express your answer in terms of the *given* parameters.

9. The function $\phi(x, y)$ is a real function of the real variables x and y. The behavior of the function for $-b \le x \le b$ and $y \ge 0$ is described by the partial differential equation

$$\frac{\partial^2 \phi(x, y)}{\partial x^2} - \frac{9}{b}\frac{\partial \phi(x, y)}{\partial x} + \frac{b}{x}\frac{\partial^2 \phi(x, y)}{\partial y^2} = 0,$$

where b is a real, positive constant. Show clearly that the method of separation of variables can be used to solve this equation. Explain your reasoning clearly. Use this method to find a solution. Require that $\phi(x, y)$ be a monotonically decreasing function of y. What constraint does this impose on the separation constant? Explain clearly your reasoning. Use the series method to obtain a solution to the ordinary differential equation in x. Require that $\phi(0, y) = 0$. What constraint does this impose on your solution? Explain clearly. With the boundary condition $\phi(0, y) = 0$, require that the x-dependent factor be a fourth-degree polynomial in x. State clearly the constraint this imposes on the separation constant. Now require that for $x = b$ and $y = 0$ the function $\phi(x, y)$ have the value λ, that is $\phi(b, 0) = \lambda$. Write down explicitly the complete expression for $\phi(x, y)$ that satisfies all of the stated conditions, namely

- $\phi(x, y)$ decreases monotonically with increasing y,
- $\phi(0, y) = 0$,
- $\phi(x, 0)$ is a fourth-degree polynomial in x,
- $\phi(b, 0) = \lambda$.

(Note: Your original separation constant should not appear explicitly as a parameter in your final result.)

10. Use the result in Eq. (6.37) to show explicitly that the electric field $\mathbf{E}$ far from the conducting sphere approaches $\mathbf{E}_0$.

    HINT: By means of a vector diagram, show that the unit vectors $\mathbf{k}$, $\hat{\mathbf{r}}$, and $\hat{\boldsymbol{\theta}}$ are related by

    $$\mathbf{k} = \hat{\mathbf{r}} \cos\theta - \hat{\boldsymbol{\theta}} \sin\theta.$$

11. Show that the electric field given in Eq. (6.37) satisfies the boundary condition at the surface of the conducting sphere.

12. At a given point in space, the electric field $\mathbf{E}$ due to an electric charge q is given by

    $$\mathbf{E} = \frac{kq}{r^3}\mathbf{r},$$

    where $\mathbf{r}$ represents the position of the field point relative to the point charge q. Show by direct differentiation that

    $$\nabla \times \mathbf{E} = 0.$$

13. An electrostatic field exists in a long, narrow rectangular region of width b. The field is derivable from an electric potential U according to

    $$\mathbf{E} = -\nabla U,$$

    where U satisfies Laplace's equation,

    $$\nabla^2 U = 0.$$

    At the base of the rectangle, the electric potential has the constant value U_0. At the other end and on the two sides, the potential is zero. The geometry (in

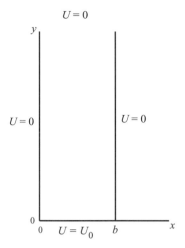

FIGURE 6.2. Electric potential for Exercise 6.13.

rectangular coordinates) is illustrated in Fig. 6.2. Because of the symmetry, there is no variation of the potential with z. Hence, $U = U(x, y)$. Use the method of separation of variables to solve Laplace's equation for $U(x, y)$. Invoke the boundary conditions at the two sides and at the top end of the region. Assume that the rectangle is so long that the upper end can be considered to be at infinity. Write $U(x, y)$ as an infinite sum of terms each of which is an explicit function of x and y multiplied by a constant coefficient (different for each term). State how the boundary condition at the bottom of the region determines these coefficients, but do not attempt to evaluate them.

14. An uncharged conducting sphere of radius R is placed in a uniform electric field. At the surface of the sphere, the electric field is given by

$$\mathbf{E} = \hat{\mathbf{r}} E_1 \cos \theta \qquad \text{for } r = R,$$

where E_1 is a constant and θ is the angle defined in Fig. 6.3. Calculate the electric potential and the electric field in the region outside the sphere. Express your results in terms of the parameters E_1 and R.

15. Solve Laplace's equation

$$\nabla^2 \phi(x, y) = 0$$

in rectangular coordinates by the method of separation of variables. The coordinates x and y are both nonnegative. Obtain the solution such that $\phi(x, y)$ is a monotonically decreasing function of y and satisfies the boundary conditions

$$\phi(0, 0) = T \qquad \text{and} \qquad \phi(R, y) = 0.$$

Express your answer in terms of the separation constant and the given parameters R and T.

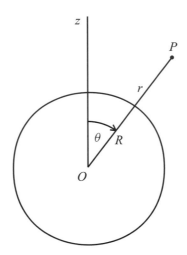

FIGURE 6.3. Conducting sphere for Exercise 6.14.

16. A scalar potential symmetric about the polar axis satisfies Laplace's equation in two dimensions

$$\nabla^2 U(r, \theta) = 0,$$

with the origin of the spherical polar coordinate system at the center of a sphere of radius R. The general solution to this partial differential equation is

$$U(r, \theta) = \sum_{n=0}^{\infty} \left(A_n r^n + \frac{B_n}{r^{n+1}} \right) P_n(\cos \theta).$$

The potential must satisfy three boundary conditions:

- $U(r, \theta) \underset{r \to \infty}{\longrightarrow} U_0 \left(1 + \dfrac{r^2}{R^2} + \dfrac{r}{R} \cos \theta - \dfrac{3r^2}{R^2} \cos^2 \theta \right)$
- $U(R, \theta)$ is independent of θ.
- $U(r, \frac{\pi}{2}) = U_0 \left(1 + \dfrac{r^2}{R^2} - \dfrac{2R}{r} - \dfrac{R^3}{r^3} \right),$

where U_0 is a constant potential. Invoke these boundary conditions to evaluate the coefficients A_n and B_n in the expansion of the potential $U(r, \theta)$. Use your results to write the potential for all points outside the sphere explicitly in the form

$$U(r, \theta) = \alpha(r) + \beta(r) \cos \theta + \gamma(r) \cos^2 \theta,$$

where α, β, and γ are functions of r. Express your result in terms of the given parameters U_0 and R.

17. Show that with the change of dependent variable $g_l(r) \equiv r R(r)$, Eq. (6.46) follows from Eq. (6.45).

18. Choose $s = -l$ in Eq. (6.48) and show that this leads to a solution to Eq. (6.45) that is irregular at $r = 0$. Show that this solution can be expressed in the form given in Eq. (6.54).

7

Eigenvalue Problems

7.1 Boundary Value Problems

In obtaining Legendre polynomials as solutions to Legendre's equation,

$$(1 - x^2)P''(x) - 2x P'(x) + \lambda P(x) = 0,$$

we have solved a typical *eigenvalue problem*. Given an equation (or set of equations) containing a parameter (here λ), we seek solutions that satisfy some special requirement (e.g., the series must converge for $x = \pm 1$). To obtain such solutions, we must choose particular values (*eigenvalues*) for the parameter. In this case,

$$\lambda = n(n + 1) \qquad \text{with} \quad n = 0, 1, 2, \ldots.$$

That is, polynomial solutions (which are required for convergence at $x = \pm 1$) arise only for certain values of λ and *not for all* values.

A differential equation with some specified auxiliary conditions for the solution to satisfy is called a *boundary value problem*. Quite often a boundary value problem leads to an eigenvalue problem. The conducting sphere in an external electric field, discussed in Chapter 6, provides an example. We required that the electric field be defined at all points in space including points on the z-axis of our chosen coordinate system. This requirement could be met only for certain values of the separation constant λ in Eq. (6.27), namely for λ equal to the product of two successive nonnegative integers (Eq. (6.29)). Eigenvalue problems appear in other forms besides differential equations as we shall see in Chapter 9.

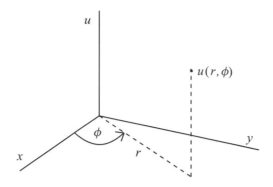

FIGURE 7.1. Displacement of drumhead relative to xy-plane.

7.2   A Vibrating Drumhead

As a further example of an eigenvalue problem, let us consider the behavior of a vibrating drumhead. A drumhead consists of a circular, elastic membrane clamped around the rim. A stroke with the drumstick displaces the drumhead by stretching it away from its equilibrium configuration. The elastic restoring force causes the drumhead to vibrate, which drives the surrounding air producing the sound that the drum makes. Let u represent the displacement of the drumhead relative to its equilibrium plane. This is illustrated in Fig. 7.1. For each point (x, y) in the horizontal equilibrium plane, there is a vertical displacement $u(x, y)$ corresponding to a point on the stretched drumhead. Because of the axial symmetry, it is convenient to represent points in the equilibrium plane in terms of the cylindrical polar coordinates (r, ϕ) shown in Fig. 7.1. Since the drumhead is vibrating, the function u also depends on time t. The complete, two-dimensional equation describing the behavior of $u(r, \phi, t)$ is[1]

$$\nabla^2 u(r, \phi, t) = \frac{1}{v^2} \frac{\partial^2}{\partial t^2} u(r, \phi, t). \tag{7.1}$$

The parameter v is the velocity of the wave propagating over the surface of the drumhead. It depends on the physical properties of the drumhead itself (e.g., the material from which it is made and how tightly it is stretched).

First, we use the method of separation of variables to write this equation as two separate equations in space and time. On setting

$$u(r, \phi, t) = g(r, \phi) f(t)$$

and substituting into Eq. (7.1), we obtain the ordinary differential equation in time,

$$\left(\frac{d^2}{dt^2} + k^2 v^2 \right) f(t) = 0 \tag{7.2}$$

---

[1]For a derivation, see A. L. Fetter and J. D. Walecka, *Theoretical Mechanics of Particles and Continua*, McGraw-Hill, New York, 1980, p. 271. See also Seaborn, *op. cit.*, p. 64

and the partial differential equation in space,

$$\frac{1}{r}\frac{\partial}{\partial r}\left(r\frac{\partial g}{\partial r}\right) + \frac{1}{r^2}\frac{\partial^2 g}{\partial \phi^2} + k^2 g = 0. \tag{7.3}$$

The separation constant is denoted by k^2. In arriving at Eq. (7.3), we have used ∇^2 in cylindrical polar coordinates as given in Appendix A.3.3. We can immediately write down the general solution to Eq. (7.2),

$$f(t) = a\cos\omega t + b\sin\omega t, \tag{7.4}$$

with $\omega \equiv kv$. Now we apply the method of separation of variables again to reduce Eq. (7.3) to two ordinary differential equations in r and ϕ separately. We set

$$g(r, \phi) = R(r)S(\phi)$$

and substitute into Eq. (7.3) to get

$$\frac{d^2 S}{d\phi^2} + \beta^2 S = 0$$

and

$$r\frac{d}{dr}\left(r\frac{dR}{dr}\right) + \left(k^2 r^2 - \beta^2\right)R = 0.$$

The separation constant here has been written as β^2. The general solution to the first of these equations is seen to be

$$S(\phi) = Ae^{i\beta\phi} + Be^{-i\beta\phi}.$$

Since (r, ϕ) and $(r, \phi + 2\pi)$ represent the same physical point in space, we must have

$$S(\phi + 2\pi) = S(\phi),$$

which requires that $\beta = $ an integer $\equiv m$.

With the change to the dimensionless independent variable, $x \equiv kr$, we can now write the second of these differential equations as

$$x^2\frac{d^2 R}{dx^2} + x\frac{dR}{dx} + \left(x^2 - m^2\right)R = 0. \tag{7.5}$$

This equation has a general solution of the form

$$R(r) = C_m J_m(kr) + D_m N_m(kr)$$

where $J_m(x)$ is regular at $x = 0$ and $N_m(x)$ is irregular at that point. For the case at hand, the displacement at the center of the drumhead ($r = 0$) cannot be infinite, so we must have $D_m = 0$ to eliminate the unphysical irregular solution. We can solve Eq. (7.5) directly by the series method to obtain the regular solution,

$$J_m(x) = \left(\frac{x}{2}\right)^m \sum_{n=0}^{\infty} \frac{(-1)^n}{n!(m+n)!}\left(\frac{x^2}{4}\right)^n.$$

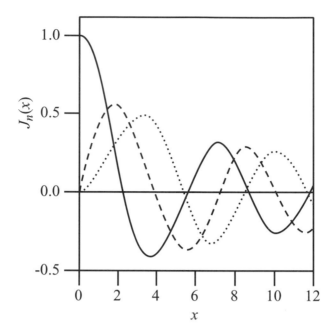

FIGURE 7.2. Bessel functions: $J_0(x)$ ——— ; $J_1(x)$ — — — ; $J_2(x)$ · · · · · · .

This function is a Bessel function of order m and Eq. (7.5) is called *Bessel's equation*. In general, the parameter m that appears in Bessel's equation and defines the order of the Bessel function need not be an integer, as we saw in Section 6.4. If m is not an integer, a more general expression for $(m + n)!$ in the series for $J_m(x)$ is required.

Bessel functions of three different orders are plotted in Fig. 7.2. This illustrates the general behavior of $J_m(x)$ as an oscillating function that decreases in amplitude as the argument x increases.

Because the drumhead is clamped at the rim, its displacement there must be zero for all values of ϕ. Taking the radius of the drumhead to be R, this boundary condition may be written as

$$J_m(kR) = 0 \qquad \text{for all } \phi.$$

Starting with the zero nearest to $x = 0$, let x_{mn} denote the position of the nth zero of $J_m(x)$. The solutions to Eq. (7.3) that satisfy the boundary conditions are

$$g_{mn}(r, \phi) = J_m\left(x_{mn}\frac{r}{R}\right)e^{im\phi}.$$

Combining this result with that of Eq. (7.4), yields solutions to Eq. (7.1) of the form

$$u(r, \phi, t) = (a\cos\omega t + b\sin\omega t)J_m\left(x_{mn}\frac{r}{R}\right)e^{im\phi},$$

where a and b are arbitrary constants. Following Eq. (7.4), we saw that the frequency ω is related to the wave number k by $\omega = vk$. We have found that k is discrete, hence the frequency may also take on only discrete allowed values according to

$$\omega_{mn} = vk_{mn} = \frac{vx_{mn}}{R}. \tag{7.6}$$

These "allowed" values of ω are called *eigenfrequencies*. Thus, for Eq. (7.1) the general solution that satisfies the required boundary conditions is

$$u(r, \phi, t) = \sum_{mn} \left(a_{mn} \cos\left(\frac{x_{mn} vt}{R}\right) + b_{mn} \sin\left(\frac{x_{mn} vt}{R}\right) \right) J_m\left(x_{mn} \frac{r}{R}\right) e^{im\phi}, \tag{7.7}$$

where the amplitudes a_{mn} and b_{mn} are determined by the initial conditions,

$$u(r, \phi, 0) \equiv p(r, \phi) \qquad \text{and} \qquad \frac{\partial u(r, \phi, t)}{\partial t}\bigg|_{t=0} \equiv q(r, \phi). \tag{7.8}$$

That is, the function $p(r, \phi)$ describes the shape of the drumhead surface at time $t = 0$ and $q(r, \phi)$ provides an analogous description of the time derivative of the function u at $t = 0$.

7.3   A Particle in a One-Dimensional Box

Now we turn to an eigenvalue problem that arises in quantum mechanics. The motion of a small particle of mass m is restricted to one dimension and to a finite range on the x-axis. The particle is in a one-dimensional box with impenetrable walls. The allowed and forbidden regions are represented pictorially in Fig. 7.3.

Information about the physical state of the particle is contained in the solution $\psi(x, t)$ of Eq. (6.1). We saw in Chapter 6 that if the potential energy V does not depend on time, the wave function takes the form

$$\psi(x, t) = u(x)e^{-iEt/\hbar},$$

where $u(x)$ satisfies Eq. (6.3). We identified the separation constant E as the total energy of the particle. For the case here, the particle has only kinetic energy in the allowed region. That is,

$$V(x) = 0 \qquad \text{for } 0 \leq x \leq L.$$

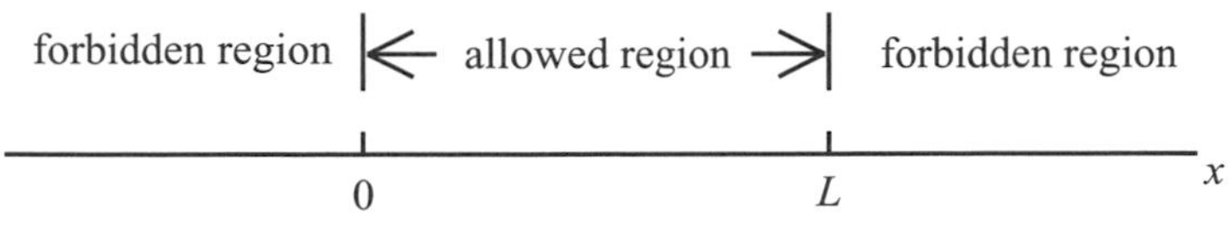

forbidden region $\Leftarrow$ allowed region $\Rightarrow$ forbidden region

FIGURE 7.3. A one-dimensional box.

In this region, Eq. (6.3) can be written

$$\frac{d^2}{dx^2}u(x) + k^2 u(x) = 0, \tag{7.9}$$

where $k \equiv \sqrt{2mE/\hbar^2}$.

The general solution to Eq. (7.9) is

$$u(x) = A \sin kx + B \cos kx. \tag{7.10}$$

There are boundary conditions that this function must satisfy. Equation (7.9) is a differential equation. Therefore, the solution $u(x)$ must be continuous everywhere including the points $x = 0$ and $x = L$. Because the particle is forbidden to exist outside the range $0 \leq x \leq L$, we expect that

$$u(x) = 0 \qquad \text{for} \quad x \leq 0, \tag{7.11}$$

and

$$u(x) = 0 \qquad \text{for} \quad x \geq L. \tag{7.12}$$

On requiring $u(x)$ to be continuous at $x = 0$, we find from Eqs. (7.10) and (7.11) that $B = 0$. Hence,

$$u(x) = A \sin kx. \tag{7.13}$$

Imposing the boundary condition at $x = L$, we see from Eqs. (7.12) and (7.13) that to avoid a trivial solution we must require $kL = n\pi$ where n is a positive integer. From the definition of $k = \sqrt{2mE/\hbar^2}$, we obtain the energy spectrum for this particle,

$$E = \frac{\pi^2 \hbar^2}{2mL^2} n^2 \qquad n = 1, 2, 3, \ldots. \tag{7.14}$$

That is, we obtain acceptable solutions to the equation of motion only for certain discrete values of the total energy of the particle. These "allowed" values given in Eq. (7.14) are the energy *eigenvalues* for this particle. The corresponding solutions to Eq. (6.3) are,

$$u_n(x) = \begin{cases} 0 & \text{for } x \leq 0 \\ A_n \sin \frac{n\pi x}{L} & \text{for } 0 \leq x \leq L \\ 0 & \text{for } x \geq L \end{cases} \quad \text{with} \quad n = 1, 2, 3, \ldots. \tag{7.15}$$

7.4   Exercises

1. Consider an elastic string stretched between two points separated by a distance L and fastened at the ends. When the string is plucked, the displacement from equilibrium of each point along the length of the string varies as a function of time. This variation manifests itself as a wave propagating along the string. At time t, the displacement from equilibrium of a point at x along the length

of the string is represented by $u(x, t)$, which satisfies the partial differential equation

$$\frac{\partial^2}{\partial x^2}u(x, t) = \frac{1}{v^2}\frac{\partial^2}{\partial t^2}u(x, t).$$

The parameter v represents the velocity at which the wave propagates along the string. Use the method of separation of variables to obtain $u(x, t)$. Choose the separation constant so as to obtain solutions that are clearly oscillatory in space and time. What are the boundary conditions (applicable for all t) on the general solution to the spatial equation? Invoke these conditions to obtain the appropriate solution to the spatial equation (apart from an overall multiplicative constant). Write the most general expression for $u(x, t)$ that satisfies these boundary conditions.

2. Write down the general solution to the differential equation

$$\frac{d^2y}{dx^2} + \lambda y = 0,$$

where λ is a real, positive constant. Find the spectrum of eigenvalues (values of λ) for which solutions to the differential equation satisfy the boundary condition: $y = 0$ for all values of x *except* in the range $0 < x < b$. By representing each eigenvalue by a short horizontal line, make an accurate sketch of the four smallest eigenvalues as illustrated in Fig. 7.4. The spaces between the lines on your drawing should be proportional to the corresponding eigenvalue differences. Label each line with the corresponding eigenvalue. Require that $y(x)$ satisfy the condition

$$\int_{-\infty}^{\infty} y^2(x)\,dx = 1,$$

and obtain an explicit expression for $y(x)$ which satisfies the boundary condition. Express your result in terms of the given parameters.

3. A small particle of mass m is confined by rigid, impenetrable walls to a finite region of the x-axis as illustrated in Fig. 7.5. For the allowed region, the potential energy is given by $V(x) = 0$ for $-a \le x \le a$. Start with Eq. (6.3) and follow an argument similar to that leading to Eq. (7.9). On

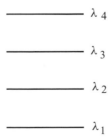

FIGURE 7.4. Eigenvalue spectrum for Exercise 7.2.

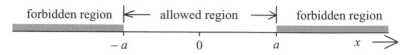

FIGURE 7.5. A particle confined by rigid walls.

the solution given in Eq. (7.10), impose boundary conditions at $x = \pm a$. Obtain the discrete energy eigenvalues and corresponding solutions to Eq. (7.9) analogous to those in Eq. (7.15). Observe the behavior of the solutions when x is replaced by $-x$. Do all of these solutions behave in the same way under this operation? Explain. The replacement of x with $-x$ is a one-dimensional example of a more general process known as the *parity* operation. Because of their special behavior under the parity operation, these solutions are said to have *definite parity*. This property of the solutions arises from the symmetry of the potential energy exhibited in Fig. 7.5. This symmetry is absent in Fig 7.3 and the corresponding solutions given in Eq. (7.15) do *not* have definite parity.

4. Solve the two-dimensional wave equation,

$$\frac{\partial^2 u(x, y, t)}{\partial x^2} + \frac{\partial^2 u(x, y, t)}{\partial y^2} = \frac{1}{v^2} \frac{\partial^2 u(x, y, t)}{\partial t^2},$$

by means of the method of separation of variables. Find the solutions that satisfy

$$\text{initial condition}: \quad \frac{\partial u(x, y, t)}{\partial t}\bigg|_{t=0} = 0,$$

$$\text{boundary conditions}: u(0, y, t) = u(a, y, t) = 0,$$

$$u(x, 0, t) = u(x, b, t) = 0.$$

where a and b are real, positive constants. Write down the most general solution that satisfies these conditions. Express your result in terms of the given parameters and the final form of your separation constants. Your solution should be an infinite series with two summation indices.

5. Impose the initial conditions given in Eqs. (7.8) on the expansion for $u(r, \phi, t)$ in Eq. (7.7). Write down explicitly the expressions that determine separately the expansion coefficients a_{mn} and b_{mn} in Eq. (7.7).

6. An elastic membrane is stretched and fastened at the edges to a rigid, rectangular frame of length a and width b. Striking the membrane sharply with a stick causes it to vibrate. The behavior of the displacement $u(x, y, t)$ of the membrane relative to its equilibrium plane is described by Eq. (7.1) expressed in rectangular coordinates x and y,

$$\nabla^2 u(x, y, t) = \frac{1}{v^2} \frac{\partial^2}{\partial t^2} u(x, y, t).$$

State clearly in mathematical terms the boundary conditions that apply here. Using the procedure in Section 7.2 as a guide, obtain an expression analogous to Eq. (7.7) that satisfies your boundary conditions. Write an explicit expression for the eigenfrequencies of the vibrational modes.

7. Two concentric conducting spherical shells are separated by a distance b. The inner shell has a radius R and carries a uniform electric potential U_1. The constant electric potential on the outer shell is U_2. The solution to Laplace's equation for the electric potential $U(r, \theta)$ at other points is given in Eq. (6.31). Consider $U(r, \theta)$ for points between the two spheres. State clearly the boundary conditions this function must satisfy. From these boundary conditions, obtain an explicit expression (in terms of U_1 and U_2) for the electric potential at points between the spheres. Use your result to obtain the electric field in this region.

8. Use the series method to find an explicit infinite series solution to Eq. (7.5).

8

Orthogonal Functions

Differential equations that arise in seeking mathematical descriptions of physical systems frequently take the form

$$\Omega u(x) = \lambda u(x)$$

as we saw in Chapter 7. Here the parameter λ is constant and the operator Ω involves differentiation with respect to x. For different values of λ, the equation has different solutions $u(x)$. We also saw in Chapter 7 that physical constraints may restrict λ to only certain values. We denote these "allowed" values by an index n. To each λ_n there corresponds a solution $u_n(x)$ to this equation. Typically, these functions obey the relation

$$\int_a^b u_m^*(x)u_n(x)\,dx = 0 \qquad \text{for } m \neq n. \tag{8.1}$$

Functions with this property are called *orthogonal functions*. They are very important and have wide application in applied mathematics. To illustrate their usefulness, we look at some specific applications, first in quantum mechanics and later in classical physics. In Section 8.6, we consider a more general treatment of orthogonal functions.

8.1   The Failure of Classical Physics

To provide a motive for considering mathematical operators and their relation to orthogonal functions, we give a brief historical sketch of the development of quantum physics. In general, *classical physics* comprises Newton's description of the mechanical behavior of matter published in 1687 and Maxwell's electrodynamics, which appeared in 1864 and provided a complete mathematical theory of light. Flaws began to appear in the structure of classical physics just around the year 1900. The difficulties arose in areas involving the interaction of these two—light (radiation) and matter. For example, classical physics failed to account for:

- The spectral distribution of radiation from an ideal radiator known as a *black body*.
- The photoelectric effect (the ejection of electrons from illuminated metal surfaces).
- The spectrum of light emitted from hot hydrogen gas.

At first, there were attempts to patch up classical physics with *ad hoc* hypotheses. These were largely unsatisfactory. What was really needed was a *new* physics. That did not come until 1925 with the invention of quantum mechanics by Werner Heisenberg and Erwin Schrödinger.

Classical physics can also be criticized on philosophical grounds. The Greeks in the fifth century B.C. speculated that one could not indefinitely subdivide matter. The view was that eventually one would come to a fundamental entity which could not be broken down into smaller pieces. Empirical evidence to support the existence of *atoms* (from a Greek word meaning *uncuttable*) came to light in the late eighteenth and early nineteenth centuries through the work of Dalton, Lavoisier, and others. However, classical physics ignores this inherent discreteness in nature and treats matter as though it were continuous. A piece of matter is considered to consist of a large number of infinitesimally small parts. If matter is inherently discrete, there must be some level on which this continuous view is no longer valid. Classical physics makes no provision for determining the scale at which such a breakdown occurs. Clearly, an *absolute* meaning of *size* is needed.[1]

8.2   Observables and Their Measurement

Science deals with *observables*. We make measurements on the physical world. To do this, the *observed* system must interact with an outside influence. If the system is such that the observation perturbs it sufficiently to affect the result of the measurement, then the system is "small" in an *absolute* sense. In such cases, a *new* theory is required. Classical physics will not work. This new theory is *quantum mechanics*.

It is a characteristic feature of small systems that an observation of the system in a given state will cause the system to make a transition to a new state. We represent this process symbolically by

observation + system in state A → system in state B.

---

[1]P. A. M. Dirac, *Principles of Quantum Mechanics*, Oxford University Press, Oxford, 1928, p. 3.

The theory must provide results corresponding to physically measurable quantities. There must be a connection between the result of an observation and the mathematical formalism. Any physical quantity that can be measured is an *observable*. In more formal language, an observable is defined[2] as a real dynamical variable whose eigenstates form a *complete set*.

For the moment, we consider the physical system to consist of a single particle of mass m whose position in space and time is denoted by the coordinates $(\mathbf{r}, t)$. The physical state of the particle is represented by a mathematical function $\Psi(\mathbf{r}, t)$. A measurement of a dynamical variable for the system in this state may perturb the system so that it goes into a *different* state as a result of the measurement. In mathematical language, this means that a mathematical operation, represented by Ω (corresponding to the dynamical variable), on $\Psi(\mathbf{r}, t)$, yields a *new* state function $\Phi(\mathbf{r}, t)$,

$$\Omega\Psi(\mathbf{r}, t) = \Phi(\mathbf{r}, t).$$

8.3   Mathematical Operators

With this motivation, we return to the notion of a mathematical *operator*. For the present, we restrict our discussion to operators that operate on functions of a single variable. The operator Ω operates on any function $u(x)$ of some variable x to produce another function $v(x)$,

$$\Omega u(x) = v(x).$$

Example 1.
Multiplication: $\Omega = x$.

$$\Omega u(x) = xu(x) = v(x).$$

Example 2.
Differentiation: $\Omega = \partial/\partial x$.

$$\Omega u(x) = \frac{\partial u(x)}{\partial x} = v(x).$$

8.4   Eigenvalue Equations[3]

An *eigenvalue equation* is a special kind of operator equation that has the general form

$$Au_n(x) = a_n u_n(x).$$

---

[2] *Ibid.* p. 37.
[3] For a more complete discussion see P. T. Matthews, *An Introduction to Quantum Mechanics*, 3rd ed., McGraw-Hill (UK), London, 1974 or Dirac, *op. cit.*.

The complex functions $u_n(x)$ are called *eigenfunctions* of the operator A and the complex numbers a_n are the corresponding *eigenvalues* of the operator A. The parameter x may represent a single variable or a set of variables. To make a physical connection with this formalism, we assert that if a measurement of the observable represented by A is made on the particle when it is in an arbitrary state, the result of the measurement will be one of the eigenvalues a_n. Furthermore, if the system is in the eigenstate $u_n(x)$, then the observation A definitely yields the result a_n. If a_n is the result one obtains in making a single physical measurement, then a_n must be a real quantity. *The eigenvalues of an operator representing an observable must be real.*

Now, suppose we want to measure A when the system is *not* in an eigenstate, but is in an arbitrary state $\psi(x)$. Sometimes the measurement will yield one value for the observable and sometimes another. To treat this case, we imagine that we have a very large number of identically prepared systems all in the same arbitrary state $\psi(x)$. The *average* value $\bar{a}$ obtained from the measurement of A on this set of identical systems must be a real quantity constructed from A and ψ in such a way that $\bar{a} = a_n$ when $\psi(x) = u_n(x)$. A number that satisfies this requirement is

$$\frac{\int u_n^*(x) A u_n(x)\, dx}{\int u_n^*(x) u_n(x)\, dx} = \frac{\int u_n^*(x) a_n u_n(x)\, dx}{\int u_n^*(x) u_n(x)\, dx} = a_n.$$

Therefore, we define the average value of the observation of A when the system is in the state $\psi(x)$ as

$$\bar{a}_\psi \equiv \frac{\int \psi^*(x) A \psi(x)\, dx}{\int \psi^*(x) \psi(x)\, dx}.$$

None of our assumptions here is affected, if any given state function is multiplied by any constant. In light of this fact, it is often convenient to *normalize* the state function $\psi(x)$ according to

$$\int \psi^*(x) \psi(x)\, dx = 1.$$

Then,

$$\bar{a}_\psi = \int \psi^*(x) A \psi(x)\, dx. \tag{8.2}$$

The reader should keep in mind that the intention here is *not* to present a definitive exposition of quantum mechanics, but merely to provide a motive for the study of operators and eigenvalue problems.[4]

---

[4]For more detail, see Matthews, *op. cit.* or Dirac, *op. cit.*

8.4.1 PHYSICAL SIGNIFICANCE OF THE STATE FUNCTION

Suppose we want to calculate the mean position (x coordinate) of the particle in the state $\psi(x)$. In this case $A = x$, and

$$\bar{x}_\psi = \int \psi^*(x) x \psi(x)\, dx = \int x |\psi(x)|^2\, dx.$$

Clearly, $|\psi(x)|^2$ is a measure of the contribution a given value of x makes to the average position $\bar{x}_\psi$. This leads us to the following statistical interpretation of $\psi(x)$:

> When the particle is in an arbitrary state $\psi(x)$, the probability that a given measurement of the particle's position will yield a value between x and $x + dx$ is
>
> $$P_\psi(x)\, dx = |\psi(x)|^2\, dx.$$

Now, let us find the probability of obtaining the result a_n when A is observed with the particle in the state $\psi(x)$. From the definition of an observable, the functions $u_n(x)$ must form a *complete* set. By completeness, we mean that any arbitrary function of x (for example, $\psi(x)$) can be written as a linear superposition of the functions $u_n(x)$. That is,

$$\psi(x) = \sum_n c_n u_n(x), \tag{8.3}$$

where the complex amplitudes c_n are independent of x. For a complete set of eigenfunctions of an operator representing a dynamical variable, we can always choose the set such that

$$\int_{-\infty}^{\infty} u_m^*(x) u_n(x)\, dx = \begin{cases} 1 & \text{for } m = n \\ 0 & \text{for } m \neq n \end{cases} \equiv \delta_{mn}. \tag{8.4}$$

The symbol δ_{mn} defined in this way is called the *Kronecker delta*.

If the eigenfunctions in Eq. (8.3) have this property, then from Eq. (8.2) we get

$$\bar{a}_\psi = \int_{-\infty}^{\infty} \psi^* A \psi\, dx = \sum_{m,n} c_m^* c_n \int_{-\infty}^{\infty} u_m^* A u_n\, dx$$

$$= \sum_{m,n} c_m^* c_n a_n \int_{-\infty}^{\infty} u_m(x)^* u_n(x)\, dx = \sum_{m,n} c_m^* c_n a_n \delta_{mn} = \sum_n |c_n|^2 a_n.$$

It is clear from this last sum that $|c_n|^2$ is a weighting factor for the value a_n in calculating the average $\bar{a}_\psi$. Thus, $|c_n|^2$ is the *probability* for obtaining the result a_n when A is observed with the particle in the state ψ.

To see how c_n is related to $u_n(x)$ and $\psi(x)$, let us multiply Eq. (8.3) by $u_m^*(x)$ and integrate over all (one-dimensional) space. This gives,

$$\int_{-\infty}^{\infty} u_m^*(x) \psi(x)\, dx = \sum_n c_n \int_{-\infty}^{\infty} u_m^*(x) u_n(x)\, dx$$

$$= \sum_n c_n \delta_{mn} = c_m.$$

So, for the mth coefficient c_m, we have

$$c_m = \int_{-\infty}^{\infty} u_m^*(x)\psi(x)\,dx. \tag{8.5}$$

Thus, the probability $P_\psi(a_n)$ for the observation A to yield the value a_n when the system is in the state $\psi(x)$ is

$$P_\psi(a_n) = \left| \int_{-\infty}^{\infty} u_n^*(x)\psi(x)\,dx \right|^2 .$$

Note that, on physical grounds, completeness is a reasonable requirement for the set of functions $u_n(x)$ to satisfy, because a completely arbitrary state $\psi(x)$ must allow for the possibility of obtaining *any* one of the eigenvalues a_n. Therefore, all possible eigenstates must be included in the set from which $\psi(x)$ is constructed.

If $\psi(x)$ is normalized in accord with Eq. (8.2), then

$$\sum_n |c_n|^2 = 1, \tag{8.6}$$

where we have used Eq. (8.4). This is a statement of the *conservation of probability*.

8.4.2   THE ENERGY EIGENVALUE EQUATION

The total nonrelativistic energy E of a small particle of mass m moving in a conservative force field is equal to the sum of its kinetic and potential energies,

$$E = \frac{\mathbf{p}^2}{2m} + V(\mathbf{r}), \tag{8.7}$$

where $\mathbf{p} = m\mathbf{v}$ is the linear momentum and $V(\mathbf{r})$ is the potential energy of the particle. Following Schrödinger, we make the transition to quantum mechanics by introducing differential operators with the correspondences,

$$E \to i\hbar \frac{\partial}{\partial t} \quad \text{and} \quad \mathbf{p} \to -i\hbar \nabla. \tag{8.8}$$

With these substitutions, Eq. (8.7) becomes an operator equation. Allowing these operators to operate on a function $\psi(\mathbf{r}, t)$ leads to the partial differential equation

$$i\hbar \frac{\partial}{\partial t} \psi(\mathbf{r}, t) = \left[\frac{-\hbar^2}{2m} \nabla^2 + V(\mathbf{r}) \right] \psi(\mathbf{r}, t), \tag{8.9}$$

which is the time-dependent Schrödinger equation.

The operator on the left-hand side of this equation depends only on time t and the operator on the right-hand side only on the space variables. Therefore, $\psi(\mathbf{r}, t)$ can be written as a product $\psi(\mathbf{r}, t) = f(t)u(\mathbf{r})$. With this substitution into Eq. (8.9), we obtain after dividing both sides by $f(t)u(\mathbf{r})$,

$$\frac{i\hbar}{f(t)} \frac{\partial}{\partial t} f(t) = \frac{1}{u(\mathbf{r})} \left[\frac{-\hbar^2}{2m} \nabla^2 + V(\mathbf{r}) \right] u(\mathbf{r}).$$

Because the left-hand side of this equation depends only on t and the right-hand side only on the position $\mathbf{r}$ with $\mathbf{r}$ and t independent variables, the two sides of the equation must be equal to a constant, which we denote by E. This provides us with an ordinary differential equation in t,

$$i\hbar \frac{d}{dt} f(t) = E f(t) \tag{8.10}$$

and a partial differential equation in the space variables,

$$\left[\frac{-\hbar^2}{2m} \nabla^2 + V(\mathbf{r}) \right] u(\mathbf{r}) = E u(\mathbf{r}). \tag{8.11}$$

The first-order differential equation, Eq. (8.10), is integrated directly to obtain $f(t) = e^{-iEt/\hbar}$. Hence, the solution to the time-dependent equation, Eq. (8.9), is,

$$\psi(\mathbf{r}, t) = e^{-iEt/\hbar} u(\mathbf{r}).$$

We see that Eq. (8.11) is in the form of an eigenvalue equation with the operator in square brackets operating on the function $u(\mathbf{r})$ (eigenfunction) to yield the same function $u(\mathbf{r})$ multiplied by a constant E (eigenvalue). In Section 6.1, we identified the separation constant E as the total energy of the particle. Hence, Eq. (8.11) is called the *energy eigenvalue equation*. The operator in square brackets in Eq. (8.11) is the total energy operator, clearly consisting of a kinetic energy term and a potential energy term.

8.5   The Quantum Harmonic Oscillator

An example of an eigenvalue problem arising from the imposition of physical constraints (boundary conditions) on solutions to a differential equation is provided by the quantum simple harmonic oscillator discussed in Chapter 6.

We required the solutions to Eq. (6.7) to be well behaved for the particle at large distances from its equilibrium position, i.e., for $|x| \rightarrow \infty$. This led to solutions to Eq. (6.7) of the form

$$u_n(x) = N_n H_n(x/b) e^{-x^2/2b^2}, \qquad \text{for } n = 0, 1, 2, 3 \ldots, \tag{8.12}$$

where $H_n(\rho)$ is a polynomial in ρ and N_n is a normalization constant. The length parameter b is defined in Eq. (6.9).

This same boundary condition also imposes a constraint on the separation constant E introduced in solving the time-dependent Schrödinger equation, Eq. (6.1), by the method of separation of variables. We identified E as the total energy of the particle. The energy eigenvalues corresponding to the solutions in Eq. (8.12) are

$$E_n = (n + \tfrac{1}{2})\hbar\omega \qquad \text{with } n = 0, 1, 2, 3, \ldots. \tag{8.13}$$

That is, the spectrum of energies for the particle is discrete. Only energies that are one-half odd-integer multiples of the fundamental unit $\hbar\omega$ are allowed. This discrete quality in nature appears clearly in the mathematics on imposing a physical constraint on the solutions to the differential equation that describes the behavior of the particle.

Normalized eigenfunctions $u_n(x)$ for the quantum harmonic oscillator are given for some low-energy eigenstates:

$$E_0 = \tfrac{1}{2}\hbar\omega: \quad u_0(x) = \left(\frac{1}{\pi b^2}\right)^{\frac{1}{4}} e^{-x^2/2b^2},$$

$$E_1 = \tfrac{3}{2}\hbar\omega: \quad u_1(x) = \left(\frac{4}{\pi b^6}\right)^{\frac{1}{4}} x e^{-x^2/2b^2},$$

$$E_2 = \tfrac{5}{2}\hbar\omega: \quad u_2(x) = \left(\frac{1}{4\pi b^2}\right)^{\frac{1}{4}} \left(\frac{2x^2}{b^2} - 1\right) e^{-x^2/2b^2},$$

$$E_3 = \tfrac{7}{2}\hbar\omega: \quad u_3(x) = \left(\frac{1}{9\pi b^2}\right)^{\frac{1}{4}} \left(\frac{2x^3}{b^3} - \frac{3x}{b}\right) e^{-x^2/2b^2},$$

$$E_4 = \tfrac{9}{2}\hbar\omega: \quad u_4(x) = \left(\frac{1}{576\pi b^2}\right)^{\frac{1}{4}} \left(\frac{4x^4}{b^4} - \frac{12x^2}{b^2} + 3\right) e^{-x^2/2b^2}.$$

Example 1.

A harmonically-bound particle oscillates about its equilibrium position. A particular state of the system is represented by the normalized function

$$\psi(x) = \left(\frac{64}{225\pi b^{14}}\right)^{\frac{1}{4}} x^3 e^{-x^2/2b^2}.$$

Calculate the probability that a measurement of the energy of the particle in the state represented by $\psi(x)$ yields each of the values $\tfrac{3}{2}\hbar\omega$ and $\tfrac{5}{2}\hbar\omega$. Calculate the average potential energy of the particle in the state $\psi(x)$.

Solution.

From the oscillator energy spectrum, Eq. (8.13), we see that $E_n = \tfrac{3}{2}\hbar\omega$ implies $n = 1$. According to our interpretation of the amplitudes c_n in Eq. (8.3), the probability for obtaining the value $\tfrac{3}{2}\hbar\omega$ is $|c_1|^2$, where c_1 is given by Eq. (8.5) with $m = 1$,

$$c_1 = \int_{-\infty}^{\infty} \left[\left(\frac{4}{\pi b^6}\right)^{\frac{1}{4}} x e^{-x^2/2b^2}\right]\left[\left(\frac{64}{225\pi b^{14}}\right)^{\frac{1}{4}} x^3 e^{-x^2/2b^2}\right] dx$$

$$= \frac{4}{\sqrt{15\pi b^5}} \int_{-\infty}^{\infty} x^4 e^{-x^2/b^2} dx.$$

Because the integrand in this last integral is an *even* function of x, we can write,

$$\int_{-\infty}^{\infty} x^4 e^{-x^2/b^2} dx = 2 \int_{0}^{\infty} x^4 e^{-x^2/b^2} dx.$$

Using the general formula,[5]

$$\int_{0}^{\infty} t^{2n} e^{-a^2 t^2} dt = \frac{(2n)!\sqrt{\pi}}{(2a)^{2n+1} n!}, \tag{8.14}$$

---

[5]See Appendix C.

where a is a real constant and n is a nonnegative integer, we get

$$\int_0^\infty x^4 e^{-x^2/b^2}\,dx = \frac{4!\sqrt{\pi}}{(2/b)^5)2!} = \frac{3b^5\sqrt{\pi}}{8}.$$

Substituting this result for the integral in the expression for c_1 above gives $c_1 = \sqrt{3/5}$. Thus, the probability that a measurement of the energy of the particle in state $\psi(x)$ yields the value $\frac{3}{2}\hbar\omega$ is $|c_1|^2 = \frac{3}{5}$.

From a similar procedure, it follows that the probability that an energy measurement in the state $\psi(x)$ yields the value $\frac{5}{2}\hbar\omega$ is given by $|c_2|^2$, where c_2 is obtained from Eq. (8.5) with $m = 2$,

$$c_2 = \frac{2}{\sqrt{15\pi b^4}} \int_{-\infty}^\infty \left(\frac{2x^5}{b^2} - x^3\right)e^{-x^2/b^2}\,dx.$$

In this case, the integrand is an *odd* function of x integrated over a symmetric interval about the origin. Therefore, the integral vanishes and $c_2 = 0$. Hence, the probability that a measurement of the energy in the state $\psi(x)$ yields $\frac{5}{2}\hbar\omega$ is zero. A measurement of the energy of the particle in this state would never result in the value $\frac{5}{2}\hbar\omega$.

To calculate the average potential energy of the particle in the state $\psi(x)$, we use Eq. (8.2) with the operator $A = V(x) = \frac{1}{2}m\omega^2 x^2$. That is,

$$\bar{V} = \int_{-\infty}^\infty \psi^*(x)\left(\tfrac{1}{2}m\omega^2 x^2\right)\psi(x)\,dx = \tfrac{1}{2}m\omega^2 \sqrt{\frac{64}{225\pi b^{14}}} \int_{-\infty}^\infty x^8 e^{-x^2/b^2}\,dx.$$

The integrand is a symmetric function of x integrated over a symmetric interval about $x = 0$, so we can use Eq. (8.14) to evaluate the integral. Thus,

$$\bar{V} = \tfrac{1}{2}m\omega^2 \sqrt{\frac{64}{225\pi b^{14}}}\, 2\,\frac{8!\sqrt{\pi}}{(2/b)^9 4!} = \frac{7m\omega^2 b^2}{4} = \tfrac{7}{4}\hbar\omega.$$

The average potential energy of the particle in the state $\psi(x)$ is $\frac{7}{4}\hbar\omega$.

8.6   Sturm-Liouville Theory[6]

Now we turn to a more general treatment of orthogonal functions. We want to consider differential equations of the form

$$\Omega v_n(x) + \lambda_n w(x)v_n(x) = 0 \tag{8.15}$$

such that the solutions have the property

$$\int_a^b v_m^*(x)v_n(x)w(x)\,dx = 0 \qquad \text{for } m \neq n. \tag{8.16}$$

---

[6]A thorough treatment of Sturm-Liouville theory is given in P. M. Morse and H. Feshbach, *Methods of Theoretical Physics*, McGraw-Hill, New York, 1953, p. 719. See also Seaborn, *op. cit.*, p. 236.

The *weighting function* $w(x)$ is real.

What constraint does the orthogonality condition, Eq. (8.16), impose on the operator Ω in Eq. (8.15)? To find out, we multiply Eq. (8.15) by $v_m^*(x)$ and integrate from a to b to get,

$$\int_a^b v_m^*(x)\Omega v_n(x)\,dx = -\lambda_n \int_a^b v_m^*(x)v_n(x)w(x)\,dx. \tag{8.17}$$

Now, change the index in Eq. (8.15) to m and take the complex conjugate to obtain

$$\left(\Omega v_m(x)\right)^* = -\lambda_m^* w(x)v_m^*(x). \tag{8.18}$$

Multiplying both sides of Eq. (8.18) by $v_n(x)$ and integrating from a to b we get

$$\int_a^b \left(\Omega v_m(x)\right)^* v_n(x)\,dx = -\lambda_m^* \int_a^b v_m^*(x)v_n(x)w(x)\,dx \tag{8.19}$$

Subtracting Eq. (8.19) from (8.17) gives

$$\int_a^b v_m^*(x)\Omega v_n(x)\,dx - \int_a^b \left(\Omega v_m(x)\right)^* v_n(x)\,dx$$

$$= (\lambda_m^* - \lambda_n) \int_a^b v_m^*(x)v_n(x)w(x)\,dx. \tag{8.20}$$

We see from Eq. (8.20) that if the functions $v_n(x)$ are to obey the orthogonality condition of Eq. (8.16), the operator Ω must have the property

$$\int_a^b v_m^*(x)\Omega v_n(x)\,dx = \int_a^b \left(\Omega v_m(x)\right)^* v_n(x)\,dx. \tag{8.21}$$

Such an operator is said to be *self-adjoint* or *hermitian*.

We shall now show that the operator

$$\Omega = \frac{d}{dx}\left[f(x)\frac{d}{dx}\right] + g(x) \tag{8.22}$$

is hermitian provided the functions $v_n(x)$ satisfy a certain condition at the limits $x = a$ and $x = b$. The functions $f(x)$ and $g(x)$ are real. We write

$$\int_a^b v_m^*(x)\Omega v_n(x)\,dx = \int_a^b v_m^*(x)\left[\frac{d}{dx}\left[f(x)\frac{dv_n(x)}{dx}\right]\right]dx$$

$$+ \int_a^b v_m^*(x)g(x)v_n(x)\,dx. \tag{8.23}$$

We take the first integral on the right-hand side of Eq. (8.23) and integrate by parts twice to get

$$\int_a^b v_m^*\left[\frac{d}{dx}\left[f\frac{dv_n}{dx}\right]\right]dx = v_m^* f\frac{dv_n}{dx}\Big|_a^b - \left(\frac{dv_m^*}{dx}\right)f v_n\Big|_a^b$$

$$+ \int_a^b \left[\frac{d}{dx}\left[f\frac{dv_m}{dx}\right]\right]^* v_n\,dx. \tag{8.24}$$

From Eq. (8.24), we make the indicated substitution on the right-hand side of Eq. (8.23) and note that if the first two terms on the right-hand side of Eq. (8.24) cancel, the operator in Eq. (8.22) is hermitian.

Using the operator Ω from Eq. (8.22), the differential equation, Eq. (8.15) can be written

$$\frac{d}{dx}\left[f(x)\frac{dv(x)}{dx} \right] + \left[g(x) + \lambda w(x) \right]v(x) = 0. \tag{8.25}$$

A differential equation of this type is known as a *Sturm-Liouville equation*. For those values of λ ($\equiv \lambda_n$), for which the corresponding solutions $v(x)$—denoted by $v_n(x)$—have the property

$$v_m^*(a)v_n'(a)f(a) = v_m^*(b)v_n'(b)f(b), \tag{8.26}$$

(i.e., for which Ω is hermitian) the solutions to a Sturm-Liouville equation are orthogonal according to Eq. (8.16).

Other differential operators besides those of the type defined in Eq. (8.22) are hermitian and have corresponding sets of orthogonal functions with a wide range of applications in applied mathematics.[7] However, differential equations in the form of Eq. (8.15) with Ω expressed as in Eq. (8.22) occur so frequently and so broadly in physics and engineering that an extensive literature has developed on these equations and their orthogonal function solutions. These well-studied functions have acquired special names such as Legendre functions, Hermite polynomials, Laguerre functions, and Bessel functions and are collectively called *special functions* or, sometimes, *higher transcendental functions*.

As a specific application, let us see how the Sturm-Liouville equation is related to the quantum harmonic oscillator problem of Section 8.5. In Section 6.2, we showed that the time-independent Schrödinger equation for the one-dimensional quantum harmonic oscillator could be written as in Eq. (6.10),

$$\frac{d^2u(\rho)}{d\rho^2} + (\lambda - \rho^2)u(\rho) = 0, \tag{8.27}$$

with the definitions

$$\lambda = \frac{2E}{\hbar\omega}, \qquad \rho = \frac{x}{b}, \qquad \text{and} \qquad b = \sqrt{\frac{\hbar}{m\omega}}.$$

According to Eq. (6.11), a solution to Eq. (8.27) may have the form

$$u(\rho) = e^{-\frac{1}{2}\rho^2} H(\rho). \tag{8.28}$$

We also saw in Section 6.2 that physically acceptable solutions of this kind are obtained only if $H(\rho)$ is a polynomial corresponding to a discrete value of λ, namely, $\lambda = 2n+1$ where n is a nonnegative integer. With this in mind, substituting for $u(\rho)$ from Eq. (8.28) in Eq. (8.27) leads to the differential equation for $H(\rho)$,

$$H_n''(\rho) - 2\rho H_n'(\rho) + 2n H_n(\rho) = 0, \tag{8.29}$$

---

[7]See, for example, the remaining sections in this chapter, and Exercise 8.15.

where we have used the index n to distinguish the polynomial solutions corresponding to the different values of λ. From Eq. (8.12), we see that the harmonic oscillator eigenfunctions are

$$u_n(\rho) = N_n H_n(\rho) e^{-\frac{1}{2}\rho^2} \qquad \text{for } n = 0, 1, 2, 3, \ldots. \qquad (8.30)$$

Now, let us write a Sturm-Liouville equation for $H_n(\rho)$,

$$\frac{d}{d\rho}\left[e^{-\rho^2} \frac{dH_n(\rho)}{d\rho} \right] + 2n\, e^{-\rho^2} H_n(\rho) = 0. \qquad (8.31)$$

In arriving at this equation from Eq. (8.25), we have set

$$\lambda = 2n, \qquad f(x) = w(x) = e^{-\rho^2}, \qquad \text{and} \qquad g(x) = 0.$$

Carrying out the differentiation in Eq. (8.31), we are led to the differential equation

$$H_n''(\rho) - 2\rho H_n'(\rho) + 2n H_n(\rho) = 0, \qquad (8.32)$$

which is identical with Eq. (8.29). That is, the polynomials $H_n(\rho)$ arising in the harmonic oscillator problem also satisfy a Sturm-Liouville equation. Note that with $f(\rho) = e^{-\rho^2}$, Eq. (8.26) is satisfied for $a = -\infty$ and $b = +\infty$. Thus, the harmonic oscillator eigenfunctions given in Eq. (8.12) are orthogonal according to,

$$\int_{-\infty}^{\infty} u_m^*(x) u_n(x)\, dx = b N_m^* N_n \int_{-\infty}^{\infty} \left(H_m(\rho) e^{-\frac{1}{2}\rho^2} \right)^* \left(H_n(\rho) e^{-\frac{1}{2}\rho^2} \right) d\rho$$

$$= b N_m^* N_n \int_{-\infty}^{\infty} H_m^*(\rho) H_n(\rho) e^{-\rho^2}\, d\rho = 0 \qquad \text{for } m \neq n,$$

$$(8.33)$$

where we have set $v_n(x) = H_n(\rho)$ and $w(x) = e^{-\rho^2}$ in applying Eq. (8.16) to the last integral in Eq. (8.33). The normalization factor N_n is chosen so that Eq. (8.4) is satisfied for $m = n$. In the standard convention, the factors N_n and the polynomials $H_n(\rho)$ are real, as seen in Section 8.4.2.[8]

8.7   The Dirac Delta Function

If the functions $u_n(x)$ form a complete set, then for any arbitrary function $f(x)$,

$$f(x) = \sum_n c_n u_n(x).$$

If the functions in the set are also orthonormal, then

$$c_n = \int_{-\infty}^{\infty} u_n^*(x) f(x)\, dx. \qquad (8.34)$$

---

[8] With a particular normalization, the real function $H_n(\rho)$ in Eq. (8.12) is called a *Hermite polynomial* of degree n.

From these two equations, we find that

$$f(x) = \int_{-\infty}^{\infty} f(x') \left[\sum_k u_k^*(x')u_k(x) \right] dx', \tag{8.35}$$

where we have reversed the order of summation and integration. We assume that $f(x')$ is a *local* function, which means that its value at $x' = x$ does not depend on its value at other points $x' \neq x$. The expression in square brackets in Eq. (8.35) can be written as,

$$\sum_k u_k^*(x')u_k(x) = \delta(x', x),$$

where $\delta(x', x)$ is a generalization to continuous indices of the Kronecker delta symbol δ_{ij}. If the index k is also continuous, then

$$f(x) = \int_{-\infty}^{\infty} c(k)u(x, k)\, dk$$

and

$$\int u^*(x', k)u(x, k)\, dk = \delta(x', x).$$

So, Eq. (8.35) becomes[9]

$$f(x) = \int_{-\infty}^{\infty} f(x')\delta(x', x)\, dx'.$$

Clearly, if $f(x)$ does not depend on values of $f(x')$ for $x' \neq x$, we have

$$\delta(x', x) = 0, \qquad \text{for } x' \neq x.$$

Taking $f(x) = 1$, we see that

$$1 = \int_{-\infty}^{\infty} \delta(x', x)\, dx'.$$

Now,

$$f(x + a) = \int_{-\infty}^{\infty} f(x' + a)\delta(x' + a, x + a)\, dx'. \tag{8.36}$$

If we consider $f(x + a)$ to be a function of x (e.g., $f(x + a) \equiv g(x)$), then

$$f(x + a) = g(x) = \int_{-\infty}^{\infty} g(x')\delta(x', x)\, dx' = \int_{-\infty}^{\infty} f(x' + a)\delta(x', x)\, dx'. \tag{8.37}$$

A comparison of Eqs. (8.36) and (8.37), shows that

$$\delta(x' + a, x + a) = \delta(x', x) \qquad \text{for any } a.$$

---

[9]This discussion through Eq. (8.39) follows E. Merzbacher, *Quantum Mechanics*, John Wiley and Sons, New York, 1961, p. 81.

Now, a is arbitrary, so let us take $a = -x$ to obtain

$$\delta(x' - x, 0) = \delta(x', x) \equiv \delta(x' - x),$$

from which we see that $\delta(x', x)$ depends only on the *difference* $x' - x$.

If $f(t)$ is any arbitrary *odd* function of t—that is, $f(-t) = -f(t)$—then $f(0) = 0$ and

$$f(0) = \int_{-\infty}^{\infty} f(t)\delta(t)\, dt = 0.$$

Clearly, $\delta(t)$ must be an *even* function of t—that is, $\delta(-t) = \delta(t)$. This quantity $\delta(x' - x)$ is known as the *Dirac delta function*[10] and has the defining properties,

$$\delta(x' - x) = 0 \qquad x' \neq x, \tag{8.38}$$

$$\int_{-\infty}^{\infty} \delta(x' - x)\, dx' = 1. \tag{8.39}$$

Note that the completeness property,

$$\sum_k u_k^*(x')u_k(x) = \delta(x' - x),$$

guarantees that $f(x)$ can be written as

$$f(x) = \sum_k c_k u_k(x).$$

This can be easily verified by substituting from Eq. (8.34) for c_k and using the properties of the delta function.

Now, $\delta(t)$ is not a function in the usual sense. It can be approximated by a narrow spike of unit area and symmetric about $t = 0$ as illustrated, for example, in Fig. 8.1.

Another approximation with the required properties is the function

$$y(t, p) = \frac{\sin pt}{\pi t}.$$

The larger p is, the sharper the spike represented by $y(t, p)$. The zeros of $y(t, p)$ occur at intervals of $t = \pi/p$ with the ones nearest to $t = 0$ at $t = \pm\pi/p$. We also have $y(0, p) = p/\pi$. So, in the limit $p \to \infty$, $y(t, p)$ becomes an infinitely sharp spike. Remarkably, in this limit the width of the spike goes to zero, the height rises to infinity, but the area under the spike *remains constant* even in the limit! If we plot $y(t, p)$ as a function of t for any p, the area under the curve is

$$\int_{-\infty}^{\infty} y(t, p)\, dt = \int_{-\infty}^{\infty} \frac{\sin pt}{\pi t}\, dt = 1, \qquad \text{for any } p.$$

---

[10] See Dirac, *op. cit.*

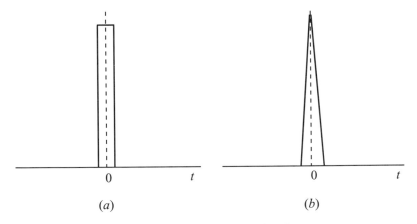

FIGURE 8.1. Sharply peaked functions.

Therefore, a suitable representation of $\delta(t)$ is

$$\delta(t) = \lim_{p \to \infty} \frac{\sin pt}{\pi t}.$$

By evaluating the following integral directly, we can write

$$\frac{1}{2\pi} \int_{-p}^{p} e^{itx} \, dx = \frac{\sin pt}{\pi t}.$$

This relation provides us with an integral representation of the delta function,

$$\delta(t) = \lim_{p \to \infty} \frac{1}{2\pi} \int_{-p}^{p} e^{itx} \, dx = \frac{1}{2\pi} \int_{-\infty}^{\infty} e^{itx} \, dx. \qquad (8.40)$$

The functions $\frac{1}{\sqrt{2\pi}} e^{ikx}$ satisfy the completeness relation, since

$$\int_{-\infty}^{\infty} \left(\frac{e^{ikx}}{\sqrt{2\pi}} \right)^{*} \left(\frac{e^{ikx'}}{\sqrt{2\pi}} \right) dk = \frac{1}{2\pi} \int_{-\infty}^{\infty} e^{i(x'-x)k} \, dk = \delta(x' - x).$$

Therefore, the elementary functions $e^{\pm ikx}$, as well as $\cos kx$ and $\sin kx$, form complete sets.

8.7.1 Another Representation

Consider the complex function,

$$y(t, p) = \frac{1}{t - \frac{i}{p}} - \frac{1}{t + \frac{i}{p}} \qquad (8.41)$$

where t and p are real. Obtaining a common denominator for the two fractions leads to the result

$$\lim_{p \to \infty} y(t, p) = \lim_{p \to \infty} \left(\frac{\frac{2i}{p}}{t^2 + \frac{1}{p^2}} \right) = 0 \qquad \text{for } t \neq 0. \qquad (8.42)$$

Now we evaluate the integral

$$\int_{-\infty}^{\infty} \lim_{p \to \infty} y(t, p) \, dt = 2i \int_{-\infty}^{\infty} \lim_{p \to \infty} \left(\frac{\frac{1}{p}}{t^2 + \frac{1}{p^2}} \right) dt. \tag{8.43}$$

With a change of the variable of integration according to $tp = \tan \theta$, the integral in Eq. (8.43) can be written as

$$\int_{-\infty}^{\infty} \lim_{p \to \infty} y(t, p) \, dt = 2i \lim_{p \to \infty} \int_{-\frac{\pi}{2}}^{\frac{\pi}{2}} \frac{\frac{1}{p^2} \sec^2 \theta \, d\theta}{\frac{1}{p^2}(1 + \tan^2 \theta)} = 2\pi i. \tag{8.44}$$

It is clear from Eqs. (8.42) and (8.44) that the defining conditions of the delta function given in Eqs. (8.38) and (8.39) are met. Hence, we have a representation of the delta function given by

$$\delta(t) = \frac{1}{2\pi i} \lim_{p \to \infty} \left(\frac{1}{t - \frac{i}{p}} - \frac{1}{t + \frac{i}{p}} \right). \tag{8.45}$$

This representation has important applications in the theory of scattering of elementary particles.

8.7.2 THE DELTA FUNCTION IN THREE DIMENSIONS

The representation of the delta function in Eq. (8.40) is easily extended to any number of dimensions. For example, in three dimensions with $\mathbf{r} = \mathbf{i}x + \mathbf{j}y + \mathbf{k}z$ we have

$$\begin{aligned}
\delta^3(\mathbf{r} - \mathbf{r}') &\equiv \delta(x - x')\delta(y - y')\delta(z - z') \\
&= \frac{1}{2\pi} \int_{-\infty}^{\infty} e^{ik_x(x-x')} \, dk_x \frac{1}{2\pi} \int_{-\infty}^{\infty} e^{ik_y(y-y')} \, dk_y \frac{1}{2\pi} \int_{-\infty}^{\infty} e^{ik_z(z-z')} \, dk_z \\
&= \frac{1}{(2\pi)^3} \int_{-\infty}^{\infty} \int_{-\infty}^{\infty} \int_{-\infty}^{\infty} e^{i[k_x(x-x')+k_y(y-y')+k_z(z-z')]} \, dk_x \, dk_y \, dk_z \\
&= \frac{1}{(2\pi)^3} \int e^{i\mathbf{k}\cdot(\mathbf{r}-\mathbf{r}')} \, d^3k.
\end{aligned}$$

In rectangular coordinates, the vector[11] $\mathbf{k} = \mathbf{i}k_x + \mathbf{j}k_y + \mathbf{k}k_z$ and the volume element $d^3k = dk_x \, dk_y \, dk_z$. However, in the last line of the above equation,

$$\delta^3(\mathbf{r} - \mathbf{r}') = \frac{1}{(2\pi)^3} \int e^{i\mathbf{k}\cdot(\mathbf{r}-\mathbf{r}')} \, d^3k,$$

it is clear that the integration can be carried out in any coordinate system. Thus, any function $f(\mathbf{r})$ of the vector $\mathbf{r}$ can be expressed as

$$f(\mathbf{r}) = \int f(\mathbf{r}')\delta^3(\mathbf{r} - \mathbf{r}') \, d^3r'.$$

---

[11]Do not confuse this vector $\mathbf{k}$ with the same symbol used for the cartesian unit vector along the z direction.

8.8   Fourier Integrals

8.8.1   FOURIER AMPLITUDES

An integral expansion of an arbitrary function $f(x)$ in terms of the complete set e^{ikx} is known as a *Fourier integral*. We write

$$f(x) = \frac{1}{\sqrt{2\pi}} \int_{-\infty}^{\infty} g(k) e^{ikx} \, dk. \tag{8.46}$$

The function $g(k)$ is called the *Fourier amplitude* of $f(x)$. Multiplying both sides of Eq. (8.46) by $e^{-ik'x}$ and integrating over the full range of x gives,

$$g(k') = \frac{1}{\sqrt{2\pi}} \int_{-\infty}^{\infty} f(x) e^{-ik'x} \, dx, \tag{8.47}$$

where we have reversed the order of integration in the double integral and invoked Eq. (8.40).

Example 4.
Suppose $f(x)$ is the rectangular pulse depicted in Fig. 8.2.

$$f(x) = \begin{cases} b & \text{for } -a < x < a \\ 0 & \text{otherwise.} \end{cases}$$

From Eq. (8.47), it follows that the Fourier amplitude for this function is

$$g(k) = \frac{1}{\sqrt{2\pi}} \int_{-a}^{a} b e^{-ikx} \, dx = \frac{2b}{\sqrt{2\pi} k} \sin ka.$$

On substituting this expression for $g(k)$ in Eq. (8.46), we obtain the Fourier integral,

$$f(x) = \frac{b}{\pi} \int_{-\infty}^{\infty} \frac{\sin ka}{k} e^{ikx} \, dk.$$

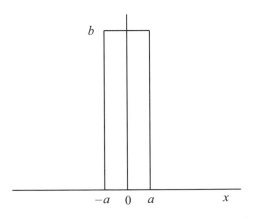

FIGURE 8.2. A rectangular pulse.

8.8.2   A PHYSICAL APPLICATION

From Eq. (8.8), we see that the momentum operator in one dimension is $p = -i\hbar \partial/\partial x$ with the corresponding momentum eigenvalue equation[12]

$$-i\hbar \frac{\partial}{\partial x} u(x, p) = p u(x, p). \tag{8.48}$$

The momentum eigenvalue is also denoted by p. By integrating this first order differential equation, we obtain the (unnormalized) eigenfunction

$$u(x, p) = e^{ipx/\hbar}.$$

We have seen that these functions form a complete set. So, for any arbitrary function $\psi(x)$, we have

$$\psi(x) = \int_{-\infty}^{\infty} c(p) e^{ipx/\hbar} \, dp. \tag{8.49}$$

In order that $\psi(x)$ be normalized to unity, we require that[13]

$$\int_{-\infty}^{\infty} |c(p)|^2 \, dp = \frac{1}{2\pi\hbar}. \tag{8.50}$$

We multiply both sides of Eq. (8.49) by $e^{-ip'x/\hbar}$ and integrate over all x to get

$$c(p) = \frac{1}{2\pi\hbar} \int_{-\infty}^{\infty} e^{-ipx/\hbar} \psi(x) \, dx.$$

In obtaining this result, we have reversed the order of integration, introduced the change of variable $y \equiv x/\hbar$, and invoked Eq. (8.40).

Example 5.

Find the wave function $\psi(x)$ which describes the behavior of a particle which has equal probability for its momentum to be any value within a finite range around the value p_0 and zero probability for its momentum to lie outside this range. This momentum distribution is given by

$$c(p) = \begin{cases} c_0 & \text{for } p_0 - \frac{\Delta p}{2} < p < p_0 + \frac{\Delta p}{2} \\ 0 & \text{otherwise.} \end{cases}$$

Solution.

From Eq. (8.49), we get

$$\psi(x) = \int_{-\infty}^{\infty} c(p) e^{ipx/\hbar} \, dp = \int_{p_0 - \frac{1}{2}\Delta p}^{p_0 + \frac{1}{2}\Delta p} c_0 e^{ipx/\hbar} \, dp$$

$$= \frac{2\hbar c_0}{x} e^{ip_0 x/\hbar} \sin\left(\frac{x\Delta p}{2\hbar}\right).$$

---

[12] A similar treatment of this problem is given in Seaborn, *op. cit.*, p. 219.
[13] See Exercise 8.30.

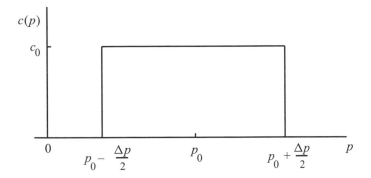

FIGURE 8.3. Momentum distribution.

Carrying out the integration in Eq. (8.50), we find that for the momentum distribution shown in Fig. 8.3, $c_0 = 1/\sqrt{2\pi\hbar\,\Delta p}$. Finally,

$$\psi(x) = \sqrt{\frac{2\hbar}{\pi\,\Delta p}}\,\frac{e^{ip_0 x/\hbar}}{x}\,\sin\!\left(\frac{x\Delta p}{2\hbar}\right).$$

In a variation of this problem, a specific wave function $\psi(x)$ for a particle is given and the task is to find the distribution of plane wave components in momentum space.

8.8.3 FOURIER INTEGRALS IN THREE DIMENSIONS

The relation

$$\delta^3(\mathbf{r} - \mathbf{r}') = \frac{1}{(2\pi)^3}\int e^{i\mathbf{k}\cdot(\mathbf{r}-\mathbf{r}')}\,d^3k$$

$$= \int \left(\frac{e^{i\mathbf{k}\cdot\mathbf{r}'}}{(2\pi)^{\frac{3}{2}}}\right)^{*}\left(\frac{e^{i\mathbf{k}\cdot\mathbf{r}}}{(2\pi)^{\frac{3}{2}}}\right)d^3k$$

$$= \int u^*(\mathbf{r}', \mathbf{k})u(\mathbf{r}, \mathbf{k})\,d^3k,$$

is clearly the completeness condition for the functions $u(\mathbf{r}, \mathbf{k}) = e^{i\mathbf{k}\cdot\mathbf{r}}/(2\pi)^{\frac{3}{2}}$. Therefore, these functions form a complete set and we can express any arbitrary function $f(\mathbf{r})$ of the vector $\mathbf{r}$ as a superposition of them according to

$$f(\mathbf{r}) = \frac{1}{(2\pi)^{\frac{3}{2}}}\int g(\mathbf{k})e^{i\mathbf{k}\cdot\mathbf{r}}\,d^3k. \tag{8.51}$$

This is the Fourier integral for the function $f(\mathbf{r})$ with the Fourier amplitude for $f(\mathbf{r})$ given by

$$g(\mathbf{k}) = \frac{1}{(2\pi)^{\frac{3}{2}}}\int f(\mathbf{r})e^{-i\mathbf{k}\cdot\mathbf{r}}\,d^3r. \tag{8.52}$$

8.9 Fourier Series

It is also true that for a *discrete* index k, the functions e^{ikx} form a complete set over an interval $a \leq x \leq a + 2\pi$. That is, for an arbitrary function $f(x)$ defined in this interval,

$$f(x) = \sum_{k=-\infty}^{\infty} c_k e^{ikx} \qquad \text{(for } a \leq x \leq a + 2\pi\text{)}. \tag{8.53}$$

To construct the coefficients c_k, we multiply both sides of Eq. (8.53) by $e^{-ik'x}$ and integrate over the interval,

$$\int_a^{a+2\pi} e^{-ik'x} f(x)\, dx = \sum_{k=-\infty}^{\infty} c_k \int_a^{a+2\pi} e^{i(k-k')x}\, dx. \tag{8.54}$$

Because of the 2π-interval and $k - k' = $ an integer, the integral on the right-hand side of Eq. (8.54) is zero if $k \neq k'$. Evaluating this integral for $k' = k$, we have

$$\int_a^{a+2\pi} e^{i(k-k')x}\, dx = 2\pi\, \delta_{kk'}. \tag{8.55}$$

Thus,

$$c_k = \frac{1}{2\pi} \int_a^{a+2\pi} e^{-ikx} f(x)\, dx. \tag{8.56}$$

The expansion in Eq. (8.53) can be written,

$$\begin{aligned}
f(x) &= c_0 + \sum_{k=1}^{\infty} c_k e^{ikx} + \sum_{k=1}^{\infty} c_{-k} e^{-ikx} \\
&= c_0 + \sum_{k=1}^{\infty} \left[c_k(\cos kx + i \sin kx) + c_{-k}(\cos kx - i \sin kx) \right] \\
&= c_0 + \sum_{k=1}^{\infty} a_k \cos kx + \sum_{k=1}^{\infty} b_k \sin kx,
\end{aligned} \tag{8.57}$$

where we have defined new coefficients according to

$$a_k = c_k + c_{-k} \qquad \text{and} \qquad b_k = i(c_k - c_{-k}).$$

In this way, we can express $f(x)$ in the interval in which it is defined as a discrete sum of sine and cosine functions. These functions have the properties,

$$\int_a^{a+2\pi} \sin mx \sin nx\, dx = \begin{cases} \pi \delta_{mn} & \text{for } m \neq 0 \\ 0 & \text{for } m = 0 \text{ or } n = 0, \end{cases} \tag{8.58}$$

$$\int_a^{a+2\pi} \cos mx \cos nx\, dx = \begin{cases} \pi \delta_{mn} & \text{for } m \neq 0 \\ 2\pi & \text{for } m = n = 0, \end{cases} \tag{8.59}$$

$$\int_a^{a+2\pi} \sin mx \cos nx\, dx = 0, \tag{8.60}$$

where m and n are integers. We use these properties to obtain the expansion coefficients in Eq. (8.57). First, multiply Eq. (8.57) by $\sin k'x$ and integrate over the interval to get

$$\int_a^{a+2\pi} f(x) \sin k'x \, dx = \sum_{k=1}^{\infty} b_k \int_a^{a+2\pi} \sin k'x \sin kx \, dx = \sum_{k=1}^{\infty} b_k \pi \delta_{kk'}.$$

Thus,

$$b_k = \frac{1}{\pi} \int_a^{a+2\pi} f(x) \sin kx \, dx.$$

Multiplying Eq. (8.57) by $\cos k'x$ and following a similar procedure yields

$$\int_a^{a+2\pi} f(x) \cos k'x \, dx = 2\pi c_0 \delta_{k'0} + \pi a_{k'} (1 - \delta_{k'0}), \tag{8.61}$$

from which we can obtain expressions for c_0 and $a_{k'}$.

Note that we can include the constant term of Eq. (8.57) in the cosine series if we define $a_0 \equiv c_0$. Finally, we write the Fourier expansion of $f(x)$ in the interval $a \le x \le a + 2\pi$ as

$$f(x) = \sum_{k=0}^{\infty} a_k \cos kx + \sum_{k=1}^{\infty} b_k \sin kx \qquad (a \le x \le a + 2\pi) \tag{8.62}$$

with

$$a_0 = \frac{1}{2\pi} \int_a^{a+2\pi} f(x) \, dx, \tag{8.63}$$

$$a_k = \frac{1}{\pi} \int_a^{a+2\pi} f(x) \cos kx \, dx \qquad (k \ge 1), \tag{8.64}$$

$$b_k = \frac{1}{\pi} \int_a^{a+2\pi} f(x) \sin kx \, dx. \tag{8.65}$$

Example 6.
Obtain the Fourier series expansion for the function $f(x)$ defined in the interval $-L < x < L$ by

$$f(x) = \lambda x \qquad (-L < x < L).$$

The function is represented graphically in Fig. 8.4.

Solution.
First, we change the independent variable to $y \equiv \pi(x/L + 1)$ so that for $-L < x < L$, we have $0 < y < 2\pi$. Then, from Eq. (8.62),

$$f(x) = \lambda L \left(\frac{y}{\pi} - 1 \right) = \sum_{k=0}^{\infty} a_k \cos ky + \sum_{k=1}^{\infty} a_k \sin ky.$$

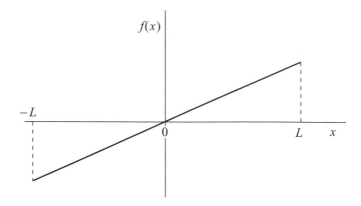

FIGURE 8.4. The function $f(x) = \lambda x$.

From Eqs. (8.63) - (8.65), the Fourier coefficients are seen to be $a_k = 0$ and

$$b_k = \frac{1}{\pi} \int_0^{2\pi} \lambda L\left(\frac{y}{\pi} - 1\right) \sin ky \, dy = -\frac{2L\lambda}{\pi k}.$$

Hence,

$$f(x) = \lambda x = -\frac{2L\lambda}{\pi} \sum_{k=1}^{\infty} \frac{1}{k} \sin \pi k\left(\frac{x}{L} + 1\right)$$

$$= -\frac{2L\lambda}{\pi} \sum_{k=1}^{\infty} \frac{(-1)^k}{k} \sin \frac{k\pi x}{L}$$

$$= \frac{2L\lambda}{\pi}\left[\sin \frac{\pi x}{L} - \frac{1}{2} \sin \frac{2\pi x}{L} + \frac{1}{3} \sin \frac{3\pi x}{L} - \cdots\right].$$

Example 7.
A cubical metal box with thin walls has edges of length s. The top and bottom faces of the box are at constant electric potential U_0. The electric potential on each of the other four faces is zero. Calculate the electrostatic potential $U(\mathbf{r})$ at any point inside the box.

Solution.
Choose a rectangular coordinate system so that the box is entirely in the first octant with one corner at the origin of the coordinate system. The bottom of the box is in the $z = 0$ plane and the top lies in the $z = s$ plane. From our work in Section 6.3, we know that $U(\mathbf{r})$ must satisfy Laplace's equation everywhere inside the box. In rectangular coordinates,

$$\nabla^2 U(\mathbf{r}) = \frac{\partial^2 U(x, y, z)}{\partial x^2} + \frac{\partial^2 U(x, y, z)}{\partial y^2} + \frac{\partial^2 U(x, y, z)}{\partial z^2} = 0. \qquad (8.66)$$

By the method of separation of variables, we may write

$$U(x, y, z) = T(x, y)Z(z).$$

Substituting this result into Eq. (8.66) and rearranging we get

$$\frac{1}{T}\left(\frac{\partial^2}{\partial x^2} + \frac{\partial^2}{\partial y^2}\right)T = -\frac{1}{Z}\frac{d^2 Z}{dz^2} = -\lambda, \tag{8.67}$$

where we have separated the z-dependent part from the (x, y)-dependent part and defined a separation constant λ. Solving Eq. (8.67) for $Z(z)$, we have

$$Z(z) = Ae^{-\sqrt{\lambda} z} + Be^{\sqrt{\lambda} z}. \tag{8.68}$$

Now, to separate the x- and y-dependent parts of Eq. (8.67), we set $T(x, y) = X(x)Y(y)$ and write

$$\frac{1}{X}\frac{d^2 X}{dx^2} = -\lambda - \frac{1}{Y}\frac{d^2 Y}{dy^2} = -\mu,$$

where we have defined a second separation constant μ. From this result, we obtain two ordinary differential equations. One of these is

$$\frac{d^2 X(x)}{dx^2} + \mu X(x) = 0,$$

with solution

$$X(x) = C\cos\sqrt{\mu}\, x + D\sin\sqrt{\mu}\, x. \tag{8.69}$$

The other equation is

$$\frac{d^2 Y(y)}{dy^2} + (\lambda - \mu)Y(y) = 0,$$

with solution

$$Y(y) = F\cos\sqrt{\lambda - \mu}\, y + G\sin\sqrt{\lambda - \mu}\, y. \tag{8.70}$$

Combining Eqs. (8.68) - (8.70), we obtain a solution to Laplace's equation, Eq. (8.66),

$$U(x, y, z) = (C\cos\sqrt{\mu}\, x + D\sin\sqrt{\mu}\, x)(F\cos\sqrt{\lambda - \mu}\, y + G\sin\sqrt{\lambda - \mu}\, y)$$
$$\times (Ae^{-\sqrt{\lambda} z} + Be^{\sqrt{\lambda} z}). \tag{8.71}$$

Applying the boundary conditions $U(0, y, z) = 0$ and $U(s, y, z) = 0$ to the solution in Eq. (8.71) leads to $C = 0$ and $\sqrt{\mu} = m\pi/s$, respectively, where m is a positive integer. Similarly, the boundary conditions $U(x, 0, z) = 0$ and $U(x, s, z) = 0$ yield $F = 0$ and $\sqrt{\lambda - \mu} = n\pi/s$, respectively, where n is a positive integer.

With these results, Eq. (8.71) can now be written,

$$U(x, y, z) = (Ae^{-\pi\sqrt{m^2+n^2}\, z/s} + Be^{\pi\sqrt{m^2+n^2}\, z/s})\sin\frac{m\pi x}{s}\sin\frac{n\pi y}{s}.$$

In this expression, we have absorbed the constants D and G in the (as yet) arbitrary constants A and B. A general solution to Laplace's equation that satisfies the four boundary conditions we have invoked so far is

$$U(\mathbf{r}) = \sum_{m=1}^{\infty} \sum_{n=1}^{\infty} (A_{mn} e^{-\pi \sqrt{m^2+n^2}\, z/s} + B_{mn} e^{\pi \sqrt{m^2+n^2}\, z/s}) \sin \frac{m\pi x}{s} \sin \frac{n\pi y}{s}.$$

To evaluate the expansion coefficients A_{mn} and B_{mn}, we have only to invoke the remaining two boundary conditions. For $z = 0$, we have

$$U(x, y, 0) = \sum_{m=1}^{\infty} \sum_{n=1}^{\infty} (A_{mn} + B_{mn}) \sin \frac{m\pi x}{s} \sin \frac{n\pi y}{s} = U_0.$$

Multiplying both sides of this equation by $\sin\left(\frac{m'\pi x}{s}\right) \sin\left(\frac{n'\pi y}{s}\right)$ and integrating over x and y from 0 to s, we get

$$\sum_{m=1}^{\infty} \sum_{n=1}^{\infty} (A_{mn} + B_{mn}) \int_0^s \sin \frac{m'\pi x}{s} \sin \frac{m\pi x}{s}\, dx \int_0^s \sin \frac{n'\pi y}{s} \sin \frac{n\pi y}{s}\, dy$$

$$= U_0 \int_0^s \sin \frac{m'\pi x}{s}\, dx \int_0^s \sin \frac{n'\pi y}{s}\, dy.$$

Using the orthogonality of the sine functions on the left-hand side of this equation and evaluating the integrals on the right-hand side yields the relation

$$A_{mn} + B_{mn} = \frac{4U_0}{mn\pi^2}\left((-1)^m - 1\right)\left((-1)^n - 1\right),$$

or

$$A_{mn} + B_{mn} = \begin{cases} \dfrac{16U_0}{mn\pi^2} & \text{for } m \text{ and } n \text{ both odd,} \\ 0 & \text{otherwise.} \end{cases}$$

Similarly, the boundary condition $U(x, y, s) = U_0$ leads to the result

$$A_{mn} e^{-\pi \sqrt{m^2+n^2}} + B_{mn} e^{\pi \sqrt{m^2+n^2}} = \begin{cases} \dfrac{16U_0}{mn\pi^2} & \text{for } m \text{ and } n \text{ both odd,} \\ 0 & \text{otherwise.} \end{cases}$$

Solving these latter two equations for A_{mn} and B_{mn}, we get

$$A_{mn} = \frac{16U_0}{mn\pi^2(1 + e^{-\pi \sqrt{m^2+n^2}})} \quad \text{and} \quad B_{mn} = \frac{16U_0}{mn\pi^2(1 + e^{\pi \sqrt{m^2+n^2}})},$$

if m and n are both odd integers. Otherwise, $A_{mn} = B_{mn} = 0$. Because m and n must both be odd, we redefine these summation indices according to $m \equiv 2p + 1$ and $n \equiv 2q + 1$ where the new indices p and q take on all nonnegative integer values. Finally, we obtain for the electric potential at any point (x, y, z) inside the box,

$$U(\mathbf{r}) = \frac{16U_0}{\pi^2} \sum_{p=0}^{\infty} \sum_{q=0}^{\infty} \left[\frac{e^{-\pi \sqrt{(2p+1)^2+(2q+1)^2}\, z/s}}{1 + e^{-\pi \sqrt{(2p+1)^2+(2q+1)^2}}} + \frac{e^{\pi \sqrt{(2p+1)^2+(2q+1)^2}\, z/s}}{1 + e^{\pi \sqrt{(2p+1)^2+(2q+1)^2}}} \right]$$

$$\times \frac{1}{(2p + 1)(2q + 1)} \sin\left[\frac{(2p + 1)\pi x}{s} \right] \sin\left[\frac{(2q + 1)\pi y}{s} \right].$$

8.10   Periodic Functions

8.10.1   FOURIER ANALYSIS OF A PERIODIC FUNCTION

Fourier series are especially useful in representing functions with the property

$$f(x + (n + 1)L) = f(x + nL) \qquad \text{for } n = 0, \pm 1, \pm 2, \cdots.$$

Functions with this property are called *periodic functions*. The parameter L is the *period* of $f(x)$. For a given value of the integer m such that x lies in the interval, $a + mL < x < a + (m + 1)L$, $f(x)$ has the same set of values as for a similar interval corresponding to a different value of the integer m.

Example 8.
A series of rectangular pulses of height h, width b, and separation b is defined by the function,

$$f(x) = \begin{cases} h & \text{for } \left(2n + \frac{1}{2}\right)b < x < \left(2n + \frac{3}{2}\right)b \\ 0 & \text{for } \left(2n + \frac{3}{2}\right)b < x < \left(2n + \frac{5}{2}\right)b, \end{cases} \tag{8.72}$$

with $n = 0, \pm 1, \pm 2, \ldots$. This function is represented graphically in Fig. 8.5. Clearly, the period of $f(x)$ is $2b$. Obtain the Fourier series for this function.

Solution.
Following the procedure in Example 6, we define a new variable $y(x)$ such that the function $f(x(y))$ is defined in the interval $0 < y < 2\pi$. The variable $y \equiv \pi(x/b - 2n)$ satisfies this requirement and the interval over which $f(x)$ is different from zero corresponds to $\pi/2 < y < 3\pi/2$. From Eqs. (8.63) - (8.65), we obtain the Fourier coefficients,

$$a_0 = \frac{1}{2\pi} \int_0^{2\pi} f(y)\, dy = \frac{1}{2\pi} \int_{\frac{\pi}{2}}^{\frac{3\pi}{2}} h\, dy = \frac{h}{2},$$

$$a_k = \frac{1}{\pi} \int_{\frac{\pi}{2}}^{\frac{3\pi}{2}} h \cos ky\, dy = \frac{h}{k\pi}\left[(-1)^k - 1\right] \sin \frac{k\pi}{2}, \tag{8.73}$$

$$b_k = \frac{1}{\pi} \int_{\frac{\pi}{2}}^{\frac{3\pi}{2}} h \sin ky\, dy = -\frac{h}{k\pi}\left[(-1)^k - 1\right] \cos \frac{k\pi}{2}. \tag{8.74}$$

From Eq. (8.73), we see that $a_k = 0$ for k equal to an even integer and

$$a_k = -\frac{2h(-1)^{\frac{k-1}{2}}}{k\pi} \qquad \text{for } k \text{ odd.}$$

Similarly, from Eq. (8.74), we get $b_k = 0$ for all k. Thus,

$$f(x) = h\left[\frac{1}{2} - \frac{2}{\pi} \sum_{m=0}^{\infty} \frac{(-1)^m}{2m + 1} \cos\left[(2m + 1)y\right]\right]$$

$$= h\left[\frac{1}{2} - \frac{2}{\pi} \sum_{m=0}^{\infty} \frac{(-1)^m}{2m + 1} \cos\left(\frac{(2m + 1)\pi x}{b}\right)\right]. \tag{8.75}$$

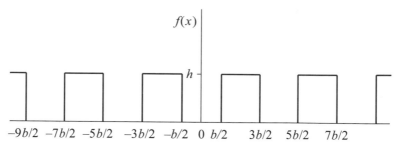

FIGURE 8.5. A series of equally spaced rectangular pulses.

Note that the function given in Eq. (8.72) is discontinuous at $x = 2nb + b/2$ and $x = 2nb + 3b/2$. For these values of x, the Fourier series in Eq. (8.75) yields $f(x) = h/2$, i.e., the arithmetic average of the values obtained from Eq. (8.72) as x approaches the position of the discontinuity from different directions.

8.10.2   A Forced Harmonic Oscillator[14]

To illustrate the power of Fourier analysis, let us reconsider the problem of the damped harmonic oscillator. In addition to the elastic restoring force and the dissipative viscous force described in Section 5.4.1, an external periodic driving force is also exerted on the particle of mass m. These three forces are represented by

- elastic force: $-m\omega_0^2 x$
- viscous force: $-2m\gamma v$
- driving force: $F(t)$

where ω_0 and γ are constant parameters and v is the velocity of the particle. The periodic driving force has the property

$$F\left(t + \frac{2\pi}{\omega}\right) = F(t). \tag{8.76}$$

Clearly, the period is $2\pi/\omega$. Following the discussion in Section 5.4.1, we arrive at the equation of motion

$$\frac{d^2 x}{dt^2} + 2\gamma \frac{dx}{dt} + \omega_0^2 x = \frac{F(t)}{m}. \tag{8.77}$$

Because the term on the right-hand side does not involve x or any of its derivatives, Eq. (8.77) is an *inhomogeneous* differential equation. We can write Eq. (8.77) as

$$\Omega x(t) = \frac{F(t)}{m} \tag{8.78}$$

---

[14] A similar treatment of this problem is given in J. B. Marion, *Classical Dynamics of Particles and Systems* 2nd ed., Academic Press, New York, 1970, p. 129.

by defining the linear operator

$$\Omega = \left[\frac{d^2}{dt^2} + 2\gamma \frac{d}{dt} + \omega_0^2 \right]. \qquad (8.79)$$

Since the force $F(t)$ is periodic, we can express it as a Fourier series,

$$F(t) = F_0 \left[a_0 + \sum_{k=1}^{\infty} \left(a_k \cos k\omega t + b_k \sin k\omega t \right) \right]. \qquad (8.80)$$

With this expression for $F(t)$, Eq. (8.78) becomes

$$\Omega x(t) = \frac{F_0}{m} \left[a_0 + \sum_{k=1}^{\infty} \left(a_k \cos k\omega t + b_k \sin k\omega t \right) \right]. \qquad (8.81)$$

The coefficients a_k and b_k are obtained from Eqs. (8.63) - (8.65) with $f(x) \equiv F(\omega t)/F_0$.[15] The terms in parentheses in the sum can be written

$$a_k \cos k\omega t + b_k \sin k\omega t = c_k \sin(k\omega t + \eta_k),$$

where we have introduced two new parameters c_k and η_k related to a_k and b_k by $a_k = c_k \sin \eta_k$ and $b_k = c_k \cos \eta_k$. We can now write Eq. (8.81) as

$$\Omega x(t) = \frac{F_0}{m} \left[a_0 + \sum_{k=1}^{\infty} c_k \sin(k\omega t + \eta_k) \right]. \qquad (8.82)$$

Before continuing with the solution of the problem with an arbitrary periodic driving force $F(t)$, let us digress to consider the simpler inhomogeneous differential equation

$$\Omega x_k(t) = \frac{d^2 x_k}{dt^2} + 2\gamma \frac{dx_k}{dt} + \omega_0^2 x_k = \frac{F_0}{m} \sin(k\omega t + \eta_k). \qquad (8.83)$$

By direct substitution in this differential equation, the reader can easily verify that the function[16]

$$x_k(t) = \frac{-F_0/m}{\sqrt{((k\omega)^2 - \omega_0^2)^2 + 4\gamma^2(k\omega)^2}} \sin(k\omega t + \eta_k + \phi_k) \qquad (8.84)$$

is a solution with the parameter ϕ_k defined by

$$\phi_k = \tan^{-1} \frac{2k\gamma\omega}{(k\omega)^2 - \omega_0^2}. \qquad (8.85)$$

The function

$$x_0(t) = \frac{F_0}{m\omega_0^2} \qquad (8.86)$$

---

[15]The factor F_0 carries the dimension of force and is introduced merely for convenience so that the Fourier coefficients a_k and b_k are dimensionless.
[16]See also Exercise 8.48.

is clearly a solution to the inhomogeneous differential equation

$$\Omega x_0(t) = \frac{d^2 x_0}{dt^2} + 2\gamma \frac{dx_0}{dt} + \omega_0^2 x_0 = \frac{F_0}{m}. \tag{8.87}$$

From Eqs. (8.83) and (8.87), we have

$$\Omega(a_0 x_0) = \frac{F_0}{m} a_0$$

$$\Omega(c_1 x_1) = \frac{F_0}{m} c_1 \sin(\omega t + \eta_1)$$

$$\Omega(c_2 x_2) = \frac{F_0}{m} c_2 \sin(2\omega t + \eta_2)$$

$$\vdots \quad = \quad \vdots \tag{8.88}$$

Setting the sum of the left-hand sides of Eqs. (8.88) equal to the sum of the right-hand sides, we get

$$\Omega \left[a_0 x_0(t) + \sum_{k=1}^{\infty} c_k x_k(t) \right] = \frac{F_0}{m} \left[a_0 + \sum_{k=1}^{\infty} c_k \sin(k\omega t + \eta_k) \right]. \tag{8.89}$$

Comparing Eq. (8.89) with Eq. (8.82), and invoking Eqs. (8.84) and (8.86), we see that

$$x(t) = a_0 x_0(t) + \sum_{k=1}^{\infty} c_k x_k(t)$$

$$= \frac{F_0}{m} \left[\frac{a_0}{\omega_0^2} - \sum_{k=1}^{\infty} c_k \frac{1}{\sqrt{\left((k\omega)^2 - \omega_0^2\right)^2 + 4\gamma^2 (k\omega)^2}} \sin(k\omega t + \eta_k + \phi_k) \right]. \tag{8.90}$$

Recalling the definitions of c_k and η_k, we write

$$c_k \sin(k\omega t + \eta_k + \phi_k)$$

$$= c_k \sin \eta_k \cos(k\omega t + \phi_k) + c_k \cos \eta_k \sin(k\omega t + \phi_k)$$

$$= a_k \cos(k\omega t + \phi_k) + b_k \sin(k\omega t + \phi_k) \tag{8.91}$$

Substituting this result into Eq. (8.90), we get for the solution to Eq. (8.77)

$$x(t) = \frac{F_0}{m} \left\{ \frac{a_0}{\omega_0^2} - \sum_{k=1}^{\infty} \frac{1}{\sqrt{\left((k\omega)^2 - \omega_0^2\right)^2 + 4\gamma^2 (k\omega)^2}} \right.$$

$$\left. \times \left[a_k \cos(k\omega t + \phi_k) + b_k \sin(k\omega t + \phi_k) \right] \right\}. \tag{8.92}$$

Here, the Fourier coefficients a_k and b_k are just those obtained in the Fourier expansion of the general periodic force $F(t)$ in Eq. (8.80). Hence, with Eq. (8.92) we have a solution $x(t)$ to Eq. (8.77).

In general, for any linear differential operator $\Omega(x)$ we can follow a similar procedure to obtain a Fourier series solution $u(x)$ to the inhomogeneous differential equation

$$\Omega(x)u(x) = s(x),$$

where the source term $s(x)$ is periodic.

8.11   Exercises

1. Show that Eq. (8.6) follows from Eqs. (8.2) and (8.4).
2. A particle of mass m is constrained to move in one dimension along the x-axis. The particle is in an arbitrary state represented by the real function $\psi(x)$, which obeys the general boundary condition

$$\psi(x) \underset{|x| \to \infty}{\to} 0.$$

   Show that the average linear momentum of the particle in this state is zero.
3. The state of a particle of mass m constrained to move along the positive x-axis is represented by the real wave function

$$\psi(x) = N(\mu x)^n e^{-\mu x} \qquad \text{with } x \geq 0,$$

   where μ is a real, positive constant and n is a positive integer. Calculate the normalization constant N such that

$$\int_{-\infty}^{\infty} \psi^*(x)\psi(x)\,dx = 1.$$

   Show that the average kinetic energy of the particle in the state $\psi(x)$ is

$$\overline{KE} = \frac{\hbar^2 \mu^2}{2m(2n - 1)}.$$

4. The state of a small particle of mass m is represented by the wave function

$$\psi(x) = \begin{cases} N(x + \mu x^2 + \mu^2 x^3)e^{-\mu x} & \text{for } x \geq 0 \\ 0 & \text{for } x < 0, \end{cases}$$

   where N and μ are real, positive constants. Find the value of N such that

$$\int_{-\infty}^{\infty} \psi^*(x)\psi(x)\,dx = 1.$$

   Calculate the average kinetic energy of the particle in the state $\psi(x)$.
5. The normalized function

$$\psi(x) = \sqrt{\frac{4}{3\sqrt{\pi}b^5}}\, x^2 e^{-x^2/2b^2}$$

can be written as a linear combination of the complete, orthonormal set of eigenfunctions $u_n(x)$ of the quantum harmonic oscillator. That is,

$$\psi(x) = \sum_{n=0}^{\infty} c_n u_n(x).$$

Calculate c_0, c_1, and c_2. Show clearly that these are sufficient to determine $\psi(x)$. Write $\psi(x)$ explicitly as a sum of the oscillator eigenfunctions $u_n(x)$.

6. The kinetic energy operator for a particle constrained to move in one dimension is

$$KE \longrightarrow \frac{-\hbar^2}{2m} \frac{d^2}{dx^2}$$

where x is the position coordinate of the particle. Suppose such a particle is in a state represented at a given time by the function

$$\psi(x) = Nx^2 e^{-\lambda x^2}$$

where N and λ are constants. Find N such that $\psi(x)$ is normalized to unity. Calculate the average value of the kinetic energy of the particle when it is in this state.

7. The energy spectrum for a simple harmonic oscillator is given by Eq. (8.13). A harmonically-bound particle is in an arbitrary state represented by the function

$$\psi(x) = \frac{4}{3\pi^{\frac{1}{4}}} \sqrt{\frac{2}{105 \, b^{11}}} x^5 e^{-x^2/2b^2}.$$

Calculate the probability that a measurement of the energy of the particle in this state yields the value $\frac{3}{2}\hbar\omega$ and the probability the measurement yields $\frac{7}{2}\hbar\omega$. Reduce each probability to a ratio of the two smallest integers. Calculate the average *potential* energy of the particle in the state $\psi(x)$. Express your result in terms of $\hbar\omega$.

8. A quantum harmonic oscillator is in a state represented by the normalized function

$$\psi(x) = \left(\frac{64}{225\pi b^{14}}\right)^{\frac{1}{4}} x^3 e^{-x^2/2b^2}.$$

Calculate the probability that a measurement of the energy of the particle in this state results in a value equal to an eigenvalue of the simple harmonic oscillator, namely, $\frac{7}{2}\hbar\omega$. Now suppose the particle is in a state represented by the normalized function,

$$\phi(x) = \left(\frac{4}{\pi b^6}\right)^{\frac{1}{4}} x e^{-x^2/2b^2}.$$

What is the probability that a measurement of the energy of the particle in this state yields the value $\frac{7}{2}\hbar\omega$?

9. Show that the harmonic oscillator eigenfunctions $u_0(x)$, $u_2(x)$, and $u_4(x)$ are orthonormal.

10. A particle of mass m is in a state represented by the function

$$\psi(x) = Nx^{2k}e^{-x^2/2b^2},$$

where N is a normalization constant, k is an integer, b is the harmonic oscillator length parameter, and x is the position of the particle relative to its equilibrium position. Show that a measurement of the total energy of the particle would never yield a value $(n+\frac{1}{2})\hbar\omega$, where ω is the angular frequency of the oscillator and n is an odd integer. Explain your reasoning clearly and completely.

11. The normalized function,

$$\psi(x) = \frac{4}{3\pi^{\frac{1}{4}}}\sqrt{\frac{2}{105\,b^{11}}}x^5 e^{-x^2/2b^2}$$

represents the state of a small particle moving under the influence of an elastic force. The function, $\psi(x)$, can also be expressed as a superposition of the harmonic oscillator wave functions $u_n(x)$ according to

$$\psi(x) = c_1 u_1(x) + c_3 u_3(x) + c_5 u_5(x).$$

Calculate c_1 and c_3. Use your results and the conservation of probability to find $u_5(x)$. Choose the phase of $u_5(x)$ so that the coefficient of the term with the highest power of x is real and positive.

12. The state of a small particle is represented by the normalized function

$$\psi(x) = \left(\frac{1}{\pi}\right)^{\frac{1}{4}}\sqrt{\frac{128}{135,135\,b^{15}}}x^7 e^{-x^2/2b^2}.$$

Calculate the mean square distance of the particle from its equilibrium position at $x = 0$ when it is in this state. Calculate the average kinetic energy of the particle in this state.

13. A small particle of mass m is in a state represented by the function

$$\psi(x) = Nx^6 e^{-x^2/2b^2},$$

where b is the oscillator length parameter and x is the displacement of the particle from equilibrium. Find N such that $\psi(x)$ is normalized to unity. Use your result to calculate the probability that in this state the total energy of the particle is greater than $\frac{11}{2}\hbar\omega$. Explain your reasoning clearly.

14. The normalized spatial wave function for a small particle confined to the x-axis is given by

$$\psi(x) = \left(\frac{1}{\pi a^2}\right)^{\frac{1}{4}}e^{-x^2/2a^2},$$

where a is a constant length parameter. (In general, a is not equal to the oscillator length parameter b.) Calculate the probability that a measurement of the total energy of the particle in this state will yield each of the values $\frac{1}{2}\hbar\omega$, $\frac{3}{2}\hbar\omega$, and $\frac{5}{2}\hbar\omega$. Use your results to obtain each of these probabilities for the special case $a = b$. Based on these last results, what do you expect the probability would be that a measurement of the energy would yield the value $(n + \frac{1}{2})\hbar\omega$ with $n > 2$, if $a = b$? Explain your reasoning clearly.

15. Set the weighting function $w(x) = 1$ in Eq. (8.15) and write down the corresponding differential equation with

$$\Omega = -i\hbar\frac{d}{dx}.$$

Require the solution $v_n(x)$ to satisfy the boundary condition

$$v_n(-a) = v_n(a).$$

Show that the operator Ω is hermitian. Find the spectrum of eigenvalues λ_n and corresponding eigenfunctions $v_n(x)$.

16. A particle of mass m is constrained to move along the positive x-axis where the potential energy of the particle is $V(x) = V_0(\frac{x}{b} - 1)$. The real, positive constant parameters V_0 and b are related by $V_0 = 3\hbar^2/5mb^2$. Calculate the average total energy of the particle when it is in a state represented by the (normalized) wave function,

$$\psi(x) = \begin{cases} 0 & \text{for } x < 0 \\ \dfrac{2}{\sqrt{3b^5}} x^2 e^{-x/b} & \text{for } x \geq 0. \end{cases}$$

Express your answer in terms of V_0.

17. A particle of mass m is confined to a one-dimensional box with impenetrable walls at $x = 0$ and at $x = L$. The energy spectrum for this particle is given by

$$E_n = \frac{\pi^2\hbar^2 n^2}{2mL^2} \qquad \text{with } n = 1, 2, 3, \ldots.$$

The corresponding normalized eigenfunctions are

$$u_n(x) = \begin{cases} \sqrt{\dfrac{2}{L}} \sin\dfrac{n\pi x}{L} & \text{for } 0 \leq x \leq L \\ 0 & \text{otherwise.} \end{cases}$$

With the particle in an arbitrary state $\psi(x)$ defined by

$$\psi(x) = \begin{cases} N \sin\dfrac{2\pi x}{L} \cos\dfrac{\pi x}{L} & \text{for } 0 \leq x \leq L \\ 0 & \text{otherwise,} \end{cases}$$

calculate the normalization constant N such that $\psi(x)$ is normalized to unity. Calculate the probability that a measurement of the total energy of the particle in the state $\psi(x)$ is $\pi^2\hbar^2/2mL^2$.

18. A particle of mass m is confined to a thin, rigid-walled, spherical shell of radius b. The energy eigenfunction corresponding to the lowest energy eigenvalue, $E_0 = \pi^2\hbar^2/2mb^2$, is

$$u_0(\mathbf{r}) = \begin{cases} \dfrac{N}{r} \sin\dfrac{\pi r}{b} & r \leq b \\ 0 & r > b. \end{cases}$$

Assume that $u_0(\mathbf{r})$ is normalized to unity and calculate the normalization constant N. (Remember that this is a problem in three dimensions.) Calculate the mean distance of the particle from the center of the sphere when it is in the state represented by $u_0(\mathbf{r})$. When the particle is in an arbitrary state (not an eigenstate) represented by the normalized function

$$\psi(\mathbf{r}) = \begin{cases} \sqrt{\dfrac{15}{2\pi b^3}}\left(1 - \dfrac{r}{b}\right) & r \le b \\ 0 & r > b, \end{cases}$$

find the probability that it has energy $\pi^2\hbar^2/2mb^2$.

19. In the simplest approximation of the hydrogen atom, the electron is bound to the proton by the electrostatic Coulomb attraction. In the lowest energy eigenstate (ground state), the energy of this system is E_0. The corresponding eigenfunction is

$$u(\mathbf{r}) = Ne^{-r/a}$$

where N is a normalization constant and a is a constant length parameter. Calculate N. (Do not ignore the angular variables.) In terms of probability, explain clearly and completely your reasoning in arriving at your answer. Calculate the average distance between the electron and the proton when the system is in the state with energy E_0. Suppose the electron is in a state (*not* an eigenstate) represented by

$$\psi(\mathbf{r}) = Ke^{-r/b}$$

where K is a normalization constant and b is a constant. Calculate K. When the system is described by $\psi(\mathbf{r})$, the probability that it has energy E_0 is 27/64. Find the length b in terms of a. There are *two* possible values of b. (The parameters a and E_0 are not arbitrary but are fixed by the values of the electron's mass and electric charge, the Coulomb constant, and Planck's constant. The actual values of a and E_0 are not important in this problem, but the fact that they are unique is important.)

20. In a model of the hydrogen atom that takes into account only the electric interaction of the electron and the proton, the energy spectrum is

$$E_n = \frac{-E_1}{n^2} \qquad \text{with } n = 1, 2, 3, \ldots.$$

The corresponding normalized energy eigenfunctions, denoted by $u_n(\mathbf{r})$, are functions of the three space coordinates represented by the vector $\mathbf{r}$. For example, for $n = 2$ we have

$$u_2(\mathbf{r}) = \sqrt{\frac{1}{32\pi a^3}}\left(2 - \frac{r}{a}\right)e^{-r/2a},$$

where a is a constant length parameter. Suppose the atom is in a state represented by the normalized function

$$\psi(\mathbf{r}) = \sqrt{\frac{1}{7\pi a^3}} \left(1 + \frac{r}{a}\right) e^{-r/a}.$$

What is the probability that a measurement of the energy in this state will yield the value $-0.25E_1$? Explain your reasoning clearly. Calculate the average distance between the electron and the proton when the system is in the state corresponding to $\psi(\mathbf{r})$.

21. A representation of the delta function is given by

$$\delta(t) = \frac{1}{\pi} \lim_{\epsilon \to 0} \left(\frac{\epsilon}{\epsilon^2 + t^2}\right),$$

where ϵ is real and nonnegative. Verify that this is a suitable representation by showing that the expression on the right-hand side satisfies the defining relations for the delta function.

22. The vector $\mathbf{r}$ represents the position of a point in space relative to a stationary point electric charge e. The charge density for this single-particle charge distribution can be represented in terms of the three-dimensional delta function according to

$$\rho(\mathbf{r}) = e\delta^3(\mathbf{r}).$$

From Gauss's law, Eq. (3.20),

$$\nabla \cdot \mathbf{E} = \frac{1}{\epsilon_0}\rho(\mathbf{r}),$$

show that the electric field at $\mathbf{r}$ is given by

$$\mathbf{E}(\mathbf{r}) = \frac{e}{4\pi\epsilon_0 r^2}\hat{\mathbf{r}}.$$

23. Calculate the Fourier amplitude for the function

$$f(x) = \begin{cases} \sin ax & \text{for } 0 \le x \le \pi/a \\ 0 & \text{otherwise} \end{cases}$$

and show that

$$f(x) = \frac{1}{\pi} \int_0^\infty \frac{a}{a^2 - k^2}\left[\cos k(x - \frac{\pi}{a}) + \cos kx\right] dk.$$

24. Calculate the Fourier amplitude $g(k)$ for the function

$$f(x) = \begin{cases} e^{-ax} & x \ge 0 \\ 0 & x < 0 \end{cases}$$

where a is a real, positive constant. Show graphically how $f(x)$ behaves as a function of x. Write $f(x)$ explicitly as a Fourier integral.

25. Calculate the Fourier amplitude for the function $e^{-x^2/2b^2}$ and show that

$$e^{-x^2/2b^2} = \frac{b}{\sqrt{2\pi}} \int_{-\infty}^\infty e^{-\frac{1}{2}b^2 k^2} e^{ikx}\, dk.$$

HINT: Complete the square in the exponent.

26. Obtain the Fourier amplitude for the function $f(x)$ given by

$$f(x) = \begin{cases} x/b & \text{for } -b \le x \le b \\ 0 & \text{for } |x| > b, \end{cases}$$

where b is a real, positive constant. Express your result in terms of trigonometric functions. Write down explicitly the Fourier integral of $f(x)$.

27. Find the explicit Fourier integral for the function

$$f(x) = b \begin{cases} \left(\frac{x}{a} + 1\right) & -a \le x \le 0 \\ \left(-\frac{x}{a} + 1\right) & 0 \le x \le a \\ 0 & \text{otherwise,} \end{cases}$$

where a and b are real, positive constants.

28. Obtain the Fourier amplitude for the function $f(x)$ given by

$$f(x) = b \begin{cases} e^{\lambda x} & \text{for } x \le 0 \\ 0 & \text{otherwise,} \end{cases}$$

where λ is a real, positive constant. Write down explicitly the Fourier integral of $f(x)$. Show all limits clearly. On a neat, clearly-labeled plot, show the behavior of $f(x)$ as a function of x.

29. The Fourier integral of a function $f(x)$ is defined by

$$f(x) = \frac{1}{\sqrt{2\pi}} \int_{-\infty}^{\infty} g(k) e^{ikx} \, dk.$$

Find the function $f(x)$ whose Fourier amplitude is

$$g(k) = \frac{1}{\sqrt{2b}} e^{(a-ik)^2/4b},$$

where a and b are real, positive constants.

30. Show that Eq. (8.50) follows from the requirement that

$$\int_{-\infty}^{\infty} \psi^*(x)\psi(x) \, dx = 1$$

for $\psi(x)$ given in Eq. (8.49).

31. Use the method of Fourier integrals, to show that

$$\frac{1}{\pi} \int_{-\infty}^{\infty} \left[\frac{(bk)^2 - 2}{k^3} \sin kb + \frac{2b}{k^2} \cos kb \right] e^{ikx} \, dk = \begin{cases} x^2 & \text{for } -b \le x \le b \\ 0 & \text{for } |x| > b. \end{cases}$$

32. Obtain the Fourier integral representation of the function $f(x)$ defined by

$$f(x) = \begin{cases} xe^{-ax} & \text{for } x \ge 0 \\ 0 & \text{for } x < 0, \end{cases}$$

where a is a real, positive constant.

33. Show that the expression for the Fourier amplitude $g(\mathbf{k})$ given in Eq. (8.52) for the function $f(\mathbf{r})$ follows from Eq. (8.51).

34. The harmonic oscillator eigenfunctions $u_n(x)$ in Eq. (8.12) and the functions $e^{ikx}/\sqrt{2\pi}$ represent complete sets of functions of x. Any arbitrary function of x can be expressed as a superposition of either set. For example, the function $e^{-x^2/2a^2}$ can be written

$$e^{-x^2/2a^2} = \sum_{n=0}^{\infty} c_n u_n(x) = \frac{1}{\sqrt{2\pi}} \int_{-\infty}^{\infty} g(k)e^{ikx}\, dk.$$

The first expression is a *discrete* sum (discrete index n) and the second is a *continuous* sum (continuous index k). Write down explicit integral expressions for the expansion coefficients c_n and $g(k)$.

35. The electrostatic potential $U(x, y)$ for the electrostatic field in Exercise 6.13 has the form

$$U(x, y) = \sum_{k=1}^{\infty} B_k \sin\frac{k\pi x}{b} e^{-k\pi y/b}.$$

This expression clearly satisfies three of the four boundary conditions given in the problem, namely

$$U(0, y) = U(b, y) = U(x, \infty) = 0.$$

Calculate the coefficients B_k such that the fourth boundary condition

$$U(x, 0) = U_0 \qquad \text{for}\quad 0 < x < b$$

is satisfied. Check your result for the particular value $x = b/2$, i.e., show that

$$U\left(\frac{b}{2}, 0\right) = U_0.$$

36. Construct the Fourier series for the function

$$f(x) = \begin{cases} \frac{h}{a}(x - b + a) & \text{for } b - a \leq x \leq b \\ -\frac{h}{a}(x - b - a) & \text{for } b \leq x \leq b + a \end{cases}$$

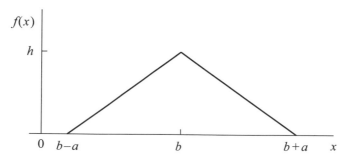

FIGURE 8.6. The function $f(x)$ for Exercise 8.36.

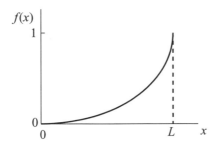

FIGURE 8.7. The function $f(x) = x^2/L^2$ for Exercise 8.37.

as shown in Fig. 8.6. To calculate the Fourier coefficients, it is convenient to define a new independent variable according to

$$y = \frac{\pi}{a}(x - b + a),$$

and rewrite $f(x)$ in terms of this variable. Verify that your expansion gives the correct values of $f(x)$ at the following values of x: $b - a$, $b - a/2$, b, $b + a/2$, and $b + a$.

37. Obtain the Fourier series representing the function $f(x)$ defined in the interval $0 < x < L$ by

$$f(x) = \frac{x^2}{L^2} \qquad 0 < x < L$$

and depicted in Fig. 8.7. Check your result for $x = L/4$ and $x = L/2$.

38. The potential energy $V(x)$ of a particle constrained to move along the x-axis in the interval $0 \le x \le L$ is

$$V(x) = \mu x^2(L - x) \qquad 0 \le x \le L.$$

Find the Fourier series representing this potential energy. Check your result for $x = 0$, $L/4$, $L/2$, and L. Find the values of x for which the particle is locally in equilibrium. At which, if any, of these positions is the equilibrium stable? Explain.

39. The elastic string under tension in Exercise 7.1 is displaced from equilibrium and released so that it vibrates in a plane. The string is fastened at the end points $x = 0$ and $x = L$. The displacement from equilibrium at time $t \ge 0$ of a point on the string at position x with $0 \le x \le L$ is represented by the function $u(x, t)$ described by the differential equation

$$\frac{\partial^2 u(x, t)}{\partial x^2} - \frac{1}{v^2}\frac{\partial^2 u(x, t)}{\partial t^2} = 0.$$

Obtain the general solution to this differential equation. Require that your solution satisfy the conditions

- Boundary conditions:

$$u(0, t) = u(L, t) = 0 \qquad \text{(fixed end points)},$$

- Initial conditions:

$$u(x, 0) = f(x)$$
$$\left. \frac{\partial u(x, t)}{\partial t} \right|_{t=0} = 0,$$

to get a solution of the form

$$u(x, t) = \sum_{n=1}^{\infty} A_n \sin \frac{n\pi x}{L} \cos \frac{n\pi vt}{L}.$$

Obtain an integral expression for the coefficients A_n. Suppose the string is pulled a distance h straight up at its midpoint as illustrated in Fig. 8.8 and released at time $t = 0$. From the information in Fig. 8.8, write the equation for $f(x)$ and use it to calculate an explicit expression for the coefficients A_n. Now write the complete series solution for $u(x, t)$ that satisfies all the given conditions. Check your result for $u(0, 0)$, $u(L/2, 0)$, and $u(L, 0)$.

40. The function

$$f(x) = \begin{cases} h - x & \text{for } 0 < x \le a \\ h + x - 2a & \text{for } a \le ax < 2a \end{cases}$$

is represented graphically in Fig. 8.9 for the case with $a < h$. Construct the Fourier series for this function valid in the interval $0 < x < 2a$.

41. A rectangular coordinate system is chosen such that in a given region of space the electrostatic potential $U(\mathbf{r})$ depends only on the x and y coordinates. That is,

$$U(\mathbf{r}) = U(x, y).$$

The region is bounded by the lines $x = 0$, $x = a$, $y = 0$, and $y = b$. The boundary conditions are:

- $U(x, 0) = 0$,
- $U(x, b) = 0$,
- $U(0, y) = U_0$,
- $\left(\frac{\partial U}{\partial x} \right)_{x=a} = 0.$

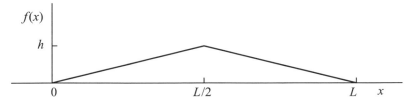

FIGURE 8.8. Initial configuration of stretched string in Exercise 8.39.

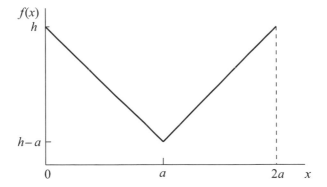

FIGURE 8.9. The function $f(x)$ in Exercise 8.40.

Obtain the unique series representation for the potential $U(x, y)$ that satisfies these boundary conditions. Express your result in terms of the given parameters a, b, and U_0.

42. A uniform static electric field $\mathbf{E}_0$ exists in a large region of space. A very long uncharged circular metal cylinder of radius R is placed in this region such that the axis of the cylinder is perpendicular to $\mathbf{E}_0$. Set up Laplace's equation for the electric potential $U(\mathbf{r})$ in cylindrical coordinates. Orient the coordinate system so as to eliminate the z-dependence of $U(\mathbf{r})$. That is, $U(\mathbf{r}) = U(r, \phi)$. Write down the boundary conditions at the surface of the cylinder and at large distances from the cylinder. Solve Laplace's equation, invoke the boundary conditions, and obtain an expression for the electric field $\mathbf{E}(r, \phi)$ at all points outside the cylinder.

43. Verify that the function defined by

$$f(x) = \sin^2 x \quad \text{for} \quad m\pi \le x \le (m+1)\pi \quad \text{with} \quad m = 0, \pm 1, \pm 2, \cdots$$

is periodic with a period π. Obtain the Fourier series for this function.

44. Plot the function

$$f(x) = \frac{x}{L}\left(\frac{x}{L} - 1\right) \quad \text{for} \ 0 \le x \le L.$$

Construct the Fourier series for this function.

45. Make an accurate plot of the function

$$f(x) = \begin{cases} h & \text{for } (4m-1)a < x < (4m+1)a \\ 0 & \text{for } (4m+1)a < x < (4m+3)a \end{cases}$$

where $m = 0, \pm 1, \pm 2, \ldots$ and verify that the function is periodic. What is the period? Does the function exhibit any symmetry? Explain. Obtain the Fourier series for this function. Are any symmetry properties of the function reflected in your result? Explain.

46. A particle is subject to a periodic, time-dependent impulsive force $F(t)$ given by

$$F(t) = \begin{cases} F_0 & \text{for } 4n\tau < t < (4n+1)\tau \\ 0 & \text{for } (4n+1)\tau < t < 4(n+1)\tau \end{cases}$$

where τ is a constant parameter and n is an integer. Make a plot showing the behavior of this force as a function of time. Construct the Fourier series representing this force.

47. The time dependence of an impulsive force $F(t)$ is

$$F(t) = \begin{cases} F_0 \cos \omega t & \text{for } \frac{(4n-1)\pi}{2\omega} < t < \frac{(4n+1)\pi}{2\omega} \\ 0 & \text{for } \frac{(4n+1)\pi}{2\omega} < t < \frac{(4n+3)\pi}{2\omega} \end{cases}$$

where the parameter ω is constant and n is an integer. Make a careful plot showing the time dependence of this force. What is the period? Construct the Fourier series for $F(t)$.

48. The equation of motion for a damped harmonic oscillator driven by a sinusoidally time-dependent force is an inhomogeneous differential equation,

$$\frac{d^2x}{dt^2} + 2\gamma \frac{dx}{dt} + \omega_0^2 x = \frac{F_0}{m} \sin(\omega t + \eta),$$

where ω_0, γ, F_0, ω, and η are constant parameters. The corresponding *homogeneous* equation is

$$\frac{d^2x_h}{dt^2} + 2\gamma \frac{dx_h}{dt} + \omega_0^2 x_h = 0.$$

We saw in Section 5.4.1 that the general solution to this latter equation is

$$x_h(t) = e^{-\gamma t}\left(Ae^{\sqrt{\gamma^2 - \omega_0^2}\,t} + Be^{-\sqrt{\gamma^2 - \omega_0^2}\,t}\right),$$

where A and B are arbitrary constants. If $x_h(t)$ is added to any solution to the inhomogeneous equation, the result is still a solution to the inhomogeneous equation. Because of the exponential damping factor $e^{-\gamma t}$, the $x_h(t)$ part of the solution will after some time become negligible and the solution to the inhomogeneous equation will be dominated by the remaining part, called the *steady-state* solution. Therefore, we expect the motion after some time will be determined by the sinusoidal driving force. With this in mind, we seek an oscillating steady-state solution that has the same frequency ω as the driving force, that is

$$x(t) = R \sin(\omega t + \theta).$$

Substitute this expression into the inhomogeneous equation and evaluate R and θ in terms of the given parameters.

49. Assume that a damped, simple harmonic oscillator is subject to the driving force $F(t)$ given in Exercise 8.47. The steady-state solution $x(t)$ to the equation of motion gives the position x of the particle at time t. Construct the Fourier series representation of $x(t)$.

50. An elastically-bound particle of mass m experiences a viscous force and a time-dependent, periodic driving force $F(t)$ given by

$$F(t) = F_0\left(1 + n - \frac{\omega t}{2\pi}\right) \qquad \text{for} \quad \frac{2n\pi}{\omega} < t < \frac{2(n+1)\pi}{\omega},$$

where n is an integer. The equation of motion for this particle is given in Eq. (8.77). Calculate the Fourier coefficients for $F(t)$ and use these to construct the steady-state solution to Eq. (8.77).

51. Show that a Sturm-Liouville equation, Eq. (8.25), with

$$f(x) = 1 - x^2, \qquad g(x) = 0, \qquad \text{and} \qquad w(x) = 1$$

reduces to Legendre's equation, Eq. (6.27),

$$(1 - x^2)P''(x) - 2x\,P'(x) + \lambda P(x) = 0.$$

To obtain polynomial solutions to this equation, we required in Section 6.3.2 that $\lambda = n(n+1)$ where n is a nonnegative integer. We used the symbol $P_n(x)$ to denote a Legendre polynomial in x of degree n. Show that with $v_n(x) = P_n(x)$ Eq. (8.26) is satisfied for $a = -1$ and $b = +1$, hence the Legendre polynomials are orthogonal on the interval $-1 \le x \le 1$. For the Legendre polynomials, write down the orthogonality relation analogous to Eq. (8.16).

52. Assume a solution of the form

$$L(x) = \sum_{k=0}^{\infty} a_k x^{k+s}$$

for the differential equation

$$x\frac{d^2 L(x)}{dx^2} + (1 - x)\frac{dL(x)}{dx} + nL(x) = 0,$$

where n is a nonnegative integer. Show that the solution is a polynomial. Label the solutions for different values of n according to $L_n(x)$. Given $f(x) = xe^{-x}$ and $v(x) = L(x)$ in Eq. (8.25), find corresponding expressions for $g(x)$ and $w(x)$ such that the Sturm-Liouville equation reduces to the above differential equation for $L(x)$. Construct $L_m(x)L_n(x)f(x)$ and identify values of a and b such that Eq. (8.26) is satisfied. This guarantees that the polynomials $L_n(x)$ are orthogonal on the interval $a \le x \le b$. Write down the explicit orthogonality relation, Eq. (8.26), for the polynomials $L_n(x)$.

9
Matrix Formulation of the Eigenvalue Problem

The earliest published version of modern quantum mechanics was Heisenberg's formulation, known as *matrix mechanics*.[1] In it, Heisenberg turned away from classical quantities that could not be measured experimentally and considered "observable magnitudes" expressible as elements of two dimensional arrays. Heisenberg's empirical rules for combining these magnitudes were identical to mathematical rules for combining matrices that were just at that time being developed by the mathematicians at Göttingen in Germany where Heisenberg was working as a physicist.

The next year, Schrödinger, following a different path, published his treatment of the behavior of matter as a wave phenomenon.[2] This formulation came to be called *wave mechanics*. Shortly thereafter, Dirac demonstrated the equivalence of these two approaches.[3] Each method has its advantages. Both are useful and find applications in different situations. The wave mechanical treatment we considered in Chapter 8 was formulated as a partial differential equation, Eq. (8.9), known as Schrödinger's equation. We turn now to some essential elements of the matrix mechanics of Heisenberg.

---

[1] W. Heisenberg, Zeitschrift für Physik **33** (1925) 879.
[2] E. Schrödinger, Annalen der Physik **79** (1926) 361.
[3] See M. Jammer, *The Conceptual Development of Quantum Mechanics*, McGraw-Hill Book Company, New York, 1966, Chapter 5 for a comprehensive review of the history of the development of quantum mechanics.

9.1   Reformulating the Eigenvalue Problem

An eigenvalue equation has the general form

$$Au_n(\mathbf{r}) = a_n u_n(\mathbf{r}). \qquad (9.1)$$

Let us assume that the set of eigenfunctions $u_n(\mathbf{r})$ is orthonormal and the eigenvalues a_n are all different. If we multiply both sides of Eq. (9.1) by $u_m^*(\mathbf{r})$ and integrate over all space, we get

$$\int u_m^*(\mathbf{r}) A u_n(\mathbf{r}) \, d^3r = a_n \delta_{mn}.$$

The left-hand side of this equation can be written as a two-dimensional array with elements

$$\int u_m^*(\mathbf{r}) A u_n(\mathbf{r}) \, d^3r \equiv A_{mn},$$

where the operator A is now represented as a matrix

$$A = \begin{pmatrix} A_{11} & A_{12} & \cdots & A_{1n} \\ A_{21} & A_{22} & \cdots & A_{2n} \\ \vdots & \vdots & \ddots & \vdots \\ A_{n1} & A_{n2} & \cdots & A_{nn} \end{pmatrix}$$

The first subscript of A_{mn} denotes the row in which the element is located and the second subscript gives the column. Thus, we can formulate an eigenvalue problem in matrix representation. A matrix A operates on a column vector X_λ to yield the same vector (an eigenvector) multiplied by a constant λ (the corresponding eigenvalue),

$$AX_\lambda = \lambda X_\lambda.$$

9.2   Systems of Linear Equations[4]

To see how this works, let's look at a well-known method for solving a set of simultaneous, linear algebraic equations for two unknowns x and y,

$$a_{11}x + a_{12}y = c_1,$$
$$a_{21}x + a_{22}y = c_2.$$

The solution is

$$x = \frac{a_{22}c_1 - a_{12}c_2}{a_{11}a_{22} - a_{12}a_{21}} \qquad\qquad y = \frac{a_{11}c_2 - a_{21}c_1}{a_{11}a_{22} - a_{12}a_{21}}. \qquad (9.2)$$

---

[4]For a more extensive discussion of matrices and systems of linear equations, see H. Margenau and G. M Murphy, *The Mathematics of Physics and Chemistry*, 2nd ed., D. Van Nostrand, Princeton, NJ, 1956, Chapter 10.

We can write the coefficients of x and y as a 2×2 array

$$\begin{pmatrix} a_{11} & a_{12} \\ a_{21} & a_{22} \end{pmatrix} \equiv A,$$

x and y as components of a 2×2 column vector

$$U = \begin{pmatrix} x \\ y \end{pmatrix} \equiv \begin{pmatrix} u_1 \\ u_2 \end{pmatrix},$$

and c_1 and c_2 as components of

$$C = \begin{pmatrix} c_1 \\ c_2 \end{pmatrix}.$$

With these definitions, the set of algebraic equations can be expressed in the compact form,

$$AU = C,$$

where A operates on the column vector U to produce the column vector C with the elements of C given by the rule

$$c_i = \sum_{j=1}^{2} a_{ij} u_j.$$

We generalize this rule to apply to an $n \times n$ matrix A operating on a column vector U with n components,

$$c_i = \sum_{j=1}^{n} a_{ij} u_j. \tag{9.3}$$

Note that the expressions for x and y (u_1 and u_2) in Eq. (9.2) have the same denominator, a number constructed from all of the elements of A. This number is the *determinant* of A. We denote it by vertical bars,

$$|A| = \begin{vmatrix} a_{11} & a_{12} \\ a_{21} & a_{22} \end{vmatrix} \equiv a_{11}a_{22} - a_{12}a_{21}.$$

To calculate determinants of three dimensions (and higher), we introduce the notion of the *cofactor* of an element. The cofactor of the element a_{ij} is given by

$$\text{cofactor of } a_{ij} = (-1)^{i+j} A^{ij}$$

where A^{ij} is the $(n-1) \times (n-1)$ determinant that remains if we strike out from $|A|$ the row and column in which a_{ij} lies. Then for any i,

$$|A| = \sum_{j=1}^{n} (-1)^{i+j} a_{ij} A^{ij}.$$

For example, if $n = 3$, we have

$$|A| - a_{11} \begin{vmatrix} a_{22} & a_{23} \\ a_{32} & a_{33} \end{vmatrix} - a_{12} \begin{vmatrix} a_{21} & a_{23} \\ a_{31} & a_{33} \end{vmatrix} + a_{13} \begin{vmatrix} a_{21} & a_{22} \\ a_{31} & a_{32} \end{vmatrix},$$

which reduces to

$$|A| = a_{11}a_{22}a_{33} + a_{12}a_{23}a_{31} + a_{13}a_{21}a_{32} - a_{11}a_{23}a_{32} - a_{12}a_{21}a_{33} - a_{13}a_{22}a_{31}.$$

The solution, Eq. (9.2), to our algebraic equation can now be written

$$x = u_1 = \frac{a_{22}c_1 - a_{12}c_2}{|A|} = \frac{\begin{vmatrix} c_1 & a_{12} \\ c_2 & a_{22} \end{vmatrix}}{|A|}$$

and

$$y = u_2 = \frac{a_{11}c_2 - a_{21}c_1}{|A|} = \frac{\begin{vmatrix} a_{11} & c_1 \\ a_{21} & c_2 \end{vmatrix}}{|A|}.$$

In generalizing this result to obtain the solution to n simultaneous equations, we are led to *Cramer's rule*.

Cramer's rule.
The jth unknown, u_j, is equal to a ratio with

$$\text{denominator} = \text{determinant } |A| \text{ of coefficients}$$

$$\text{numerator} = \text{determinant of coefficients with } j\text{th}$$
$$\text{column replaced by elements of } C$$
$$(\text{i.e., } a_{ij} \to c_i)$$

Suppose the system of equations is *homogeneous*, that is $c_i = 0$ for all i. In this case, all numerators for the u_j vanish and we obtain a trivial solution ($u_j = 0$ for all j) *unless* $|A| = 0$.

9.3   Back to the Eigenvalue Problem

Now let's return to the eigenvalue problem in matrix form,

$$AX_\lambda = \lambda X_\lambda$$

which we can rearrange as

$$(A - \lambda \mathbf{1})X_\lambda = 0,$$

where $\mathbf{1}$ is the unit matrix. Here we have a matrix $A - \lambda \mathbf{1}$ operating on the column vector X_λ to yield zero. The problem of finding the elements of X_λ is the same as looking for nontrivial solutions to a set of *homogeneous* algebraic equations by applying Cramer's rule. As we have seen, a nontrivial solution exists only if the determinant $|A - \lambda \mathbf{1}| = 0$.

Example 1.

Find the eigenvalues and eigenvectors of the matrix A,

$$A = \begin{pmatrix} 3 & 1 & 0 \\ 1 & 3 & 0 \\ 0 & 0 & 1 \end{pmatrix}.$$

Solution.

First, set the determinant $|A - \lambda\mathbf{1}| = 0$ with the result

$$|A - \lambda\mathbf{1}| = (1 - \lambda)(2 - \lambda)(4 - \lambda) = 0.$$

Clearly, this equation has three distinct roots, which we label according to

$$\lambda_1 = 1 \qquad\qquad \lambda_2 = 2 \qquad\qquad \lambda_3 = 4.$$

These are the eigenvalues of the matrix A. Now, we must find the corresponding eigenvectors that satisfy

$$AX_l = \lambda_l X_l.$$

We begin with $\lambda_1 = 1$ and write

$$(A - 1)X_1 = \begin{pmatrix} 2 & 1 & 0 \\ 1 & 2 & 0 \\ 0 & 0 & 0 \end{pmatrix} \begin{pmatrix} x_{11} \\ x_{21} \\ x_{31} \end{pmatrix} = \begin{pmatrix} 2x_{11} + x_{21} \\ x_{11} + 2x_{21} \\ 0 \end{pmatrix} = \begin{pmatrix} 0 \\ 0 \\ 0 \end{pmatrix}. \qquad (9.4)$$

Note that the second index on the element x_{ij} denotes the column vector X_j to which the element belongs. Solving Eq. (9.4) for the x_{i1}, we find that

$$x_{21} = -2x_{11} \qquad\qquad x_{11} = -2x_{21} \qquad\qquad x_{31} = \text{arbitrary.}$$

From this, it is clear that $x_{11} = x_{21} = 0$ and x_{31} is arbitrary. In a similar fashion, one can find the components of the column vectors obtained from the other two eigenvalues $\lambda_2 = 2$ and $\lambda_3 = 4$. Thus, the eigenvectors (normalized to unity) corresponding to the eigenvalues λ_i are

$$X_1 = \begin{pmatrix} 0 \\ 0 \\ 1 \end{pmatrix} \qquad X_2 = \frac{1}{\sqrt{2}} \begin{pmatrix} 1 \\ -1 \\ 0 \end{pmatrix} \qquad X_3 = \frac{1}{\sqrt{2}} \begin{pmatrix} 1 \\ 1 \\ 0 \end{pmatrix}. \qquad (9.5)$$

To see that these vectors are orthogonal, we need to establish a rule for the scalar product of two column vectors X and Y. First, we require some definitions.

Definitions.

• The *complex conjugate* of a matrix A is denoted by A^* with the matrix elements of A^* equal to the complex conjugates of the corresponding matrix elements of A, that is

$$(A^*)_{ij} = A^*_{ij}.$$

• The *transpose* of a matrix A is denoted by $\tilde{A}$ in which the rows of A become the columns of $\tilde{A}$. Thus, the matrix elements of $\tilde{A}$ are related to those of A by

$$(\tilde{A})_{ij} = A_{ji}.$$

- The *adjoint* of a matrix A is denoted by $A^\dagger$ and represents a combination of the previous two operations on A. That is, we take the complex conjugates of all the elements in A and then interchange the rows and columns of the resulting matrix to get $A^\dagger$ with elements

$$(A^\dagger)_{ij} = A^*_{ji} \tag{9.6}$$

Applying these definitions to the column matrix,

$$X = \begin{pmatrix} x_1 \\ x_2 \\ \vdots \\ x_n \end{pmatrix}$$

we see that $X^\dagger$ is a row matrix,

$$X^\dagger = (x_1^* \quad x_2^* \quad \cdots \quad x_n^*).$$

Now, we define the scalar product of two n-dimensional column vectors X and Y as

$$\text{scalar product} = X^\dagger Y = (x_1^* \quad x_2^* \quad \cdots \quad x_n^*) \begin{pmatrix} y_1 \\ y_2 \\ \vdots \\ y_n \end{pmatrix}$$

$$= x_1^* y_1 + x_2^* y_2 + \cdots + x_n^* y_n = \sum_{k=1}^{n} x_k^* y_k,$$

which is a number. If this number is zero, then X and Y are *orthogonal*. Now, if we apply this to the eigenvectors we obtained in Eq. (9.5) we see that indeed

$$X_i^\dagger X_j = 0 \qquad \text{for all } i \neq j.$$

So, for the vectors in Eq. (9.5), we have

$$X_i^\dagger X_j = \delta_{ij}.$$

That is, the vectors X_i are orthonormal.

9.3.1   EIGENVALUES FOR A 3×3 MATRIX

To illustrate the essential ideas and at the same time keep the computations relatively simple, we consider only 2×2 and 3×3 matrices. Obtaining the eigenvalues from the determinants of these matrices involves finding the roots of quadratic and cubic equations, respectively. General methods for solving these equations are reviewed in Appendix D.

Example 2.
Use the method outlined in Appendix D to find the roots of

$$x^3 - 6x^2 - 6x + 63 = 0.$$

Solution.

First, we set $x = y + 2$ to obtain Eq. (D.4) with $t = -18$ and $u = 35$. With these values for t and u, the solution to Eq. (D.5) is

$$z^3 = -8 = 8e^{i\pi}.$$

From Eq. (D.7), we have

$$z = 2e^{i(2m+1)\pi/3} \qquad \text{with} \quad m = 0, 1, 2.$$

From these values of z, we construct the corresponding values of y and x. For example, the $m = 0$ solution yields,

$$x = y + 2 = \tfrac{1}{2}(9 - i\sqrt{3}).$$

It is left as an exercise for the reader to verify that this is one of the roots by substituting it into the cubic equation for x.

9.3.2   MATRIX MULTIPLICATION

The product of two $n \times n$ matrices A and B yields a third $n \times n$ matrix C. We write this operation as

$$AB = C,$$

where the elements of C are given by the rule

$$C_{ij} = (AB)_{ij} \equiv \sum_{k=1}^{n} A_{ik} B_{kj}. \tag{9.7}$$

Note that, in general, $AB \neq BA$.

With the rule in Eq. (9.7), it is straightforward to show that

$$\widetilde{AB} = \tilde{B}\tilde{A} \qquad \text{and} \qquad (AB)^\dagger = B^\dagger A^\dagger.$$

Successive applications of the rule allow us to generalize these results to products of any number of matrices,

$$\widetilde{AB \cdots R} = \tilde{R} \cdots \tilde{B}\tilde{A} \qquad \text{and} \qquad (AB \cdots R)^\dagger = R^\dagger \cdots B^\dagger A^\dagger. \tag{9.8}$$

That is, the transpose of a product of matrices is equal to the product of the transposed matrices in reverse order. A similar statement applies to the adjoint of a product of matrices.

In quantum mechanics, it is often the matrix elements

$$A_{ij} = \int u_i^*(\mathbf{r}) A u_j(\mathbf{r}) \, d^3 r$$

that are the physically measurable quantities, e.g., amplitudes representing transitions between states of a physical system or the average value of the measurement of *another* dynamical variable B, such as

$$\bar{B} = \int \psi^*(\mathbf{r}) B \psi(\mathbf{r}) \, d^3 r = \sum_{k,l} c_k^* c_l \int u_k^*(\mathbf{r}) B u_l(\mathbf{r}) \, d^3 r \equiv \sum_{k,l} c_k^* c_l B_{kl} \cdot$$

So, the problem can now be stated as follows: *Given a matrix A, find the spectrum of eigenvalues and the corresponding eigenvectors.*

If A is a dynamical variable, and the eigenvalue λ_i is the result of a measurement of A on a given system, then λ_i must be a real quantity (i.e., $\lambda_i^* = \lambda_i$). To see the implication of this constraint, consider a square matrix A with the property $A^\dagger = A$ where the elements of $A^\dagger$ are related to those of A according to Eq. (9.6). A matrix with this property is said to be *hermitian*.

For simplicity, we shall restrict the discussion to a hermitian matrix for which the eigenvalues are all different, that is $\lambda_i \neq \lambda_j$ for $i \neq j$. Let us write down the eigenvalue equation for this matrix twice,

$$AX_i = \lambda_i X_i \qquad \text{and} \qquad AX_j = \lambda_j X_j$$

where j may or may not be equal to i. For the first of these equations, we multiply both sides from the left with the row vector $X_j^\dagger$ to obtain

$$X_j^\dagger A X_i = \lambda_i X_j^\dagger X_i. \tag{9.9}$$

We make a similar operation with $X_i^\dagger$ on the second equation to get

$$X_i^\dagger A X_j = \lambda_j X_i^\dagger X_j. \tag{9.10}$$

Now take the adjoint of Eq. (9.9) and subtract the result from Eq. (9.10). Then, invoking Eq. (9.8), we see that

$$(X_i^\dagger A X_j) - (X_i^\dagger A^\dagger X_j) = (\lambda_j - \lambda_i^*)X_i^\dagger X_j.$$

Since A is hermitian, the left-hand side of this equation is zero. Then for the right-hand side, we have

$$(\lambda_j - \lambda_i^*)X_i^\dagger X_j = 0.$$

There are two cases:

$i = j$: $(\lambda_i - \lambda_i^*)|X_i|^2 = 0$, from which we see that λ_i must be real for all i.

$i \neq j$: $(\lambda_j - \lambda_i)X_i^\dagger X_j = 0$, which requires $X_i^\dagger X_j = 0$.

This second case implies that the eigenvectors are orthogonal. Let's also assume that they are normalized to unity, i.e.,

$$X_i^\dagger X_j = \delta_{ij},$$

in which case Eq. (9.10) reduces to

$$X_i^\dagger A X_j = \lambda_i \delta_{ij}. \tag{9.11}$$

The right-hand side of Eq. (9.11) gives the elements of the matrix

$$\Lambda = \begin{pmatrix} \lambda_1 & 0 & 0 & \cdots \\ 0 & \lambda_2 & 0 & \cdots \\ 0 & 0 & \lambda_3 & \cdots \\ \vdots & \vdots & \vdots & \ddots \end{pmatrix}.$$

That is, Λ is a diagonal matrix with eigenvalues along the diagonal and zeros everywhere else.

9.3.3 DIAGONALIZATION OF A HERMITIAN MATRIX

Suppose we let X_j be the jth column of a matrix S,

$$X_j = \begin{pmatrix} x_{1j} \\ x_{2j} \\ \vdots \\ x_{ij} \\ \vdots \end{pmatrix} = \begin{pmatrix} S_{1j} \\ S_{2j} \\ \vdots \\ S_{ij} \\ \vdots \end{pmatrix}$$

or

$$S = \begin{pmatrix} S_{11} & S_{12} & \cdots & x_{1j} & \cdots \\ S_{21} & S_{22} & \cdots & x_{2j} & \cdots \\ \vdots & \vdots & \vdots & \vdots & \vdots \\ S_{i1} & S_{i2} & \cdots & x_{ij} & \cdots \\ \vdots & \vdots & \vdots & \vdots & \end{pmatrix}.$$

Clearly, $S_{ij} = x_{ij}$ is the ith element of the jth vector. For the adjoint of S, we have

$$S^\dagger = \begin{pmatrix} S_{11}^* & S_{21}^* & \cdots & S_{i1}^* & \cdots \\ S_{12}^* & S_{22}^* & \cdots & S_{i2}^* & \cdots \\ \vdots & \vdots & \vdots & \vdots & \vdots \\ x_{1j}^* & x_{2j}^* & \cdots & x_{ij}^* & \cdots \\ \vdots & \vdots & \vdots & \vdots & \end{pmatrix} = \begin{pmatrix} S_{11}^* & S_{21}^* & \cdots & S_{i1}^* & \cdots \\ S_{12}^* & S_{22}^* & \cdots & S_{i2}^* & \cdots \\ \vdots & \vdots & \vdots & \vdots & \vdots \\ S_{1j}^* & S_{2j}^* & \cdots & S_{ij}^* & \cdots \\ \vdots & \vdots & \vdots & \vdots & \end{pmatrix}$$

from which we see that $(S^\dagger)_{ji} = x_{ij}^* = S_{ij}^*$. Thus, $X_i^\dagger A X_j$ is a number that can be written as

$$X_i^\dagger A X_j = \sum_{k,l} x_{ki}^* A_{kl} x_{lj} = \sum_{k,l} (S^\dagger)_{ik} A_{kl} S_{lj} = (S^\dagger A S)_{ij} = \lambda_i \delta_{ij}.$$

So the eigenvectors of A are columns of a matrix S that will bring A to diagonal form. That is,

$$S^\dagger A S = \Lambda,$$

where Λ is a diagonal matrix.

The matrices A and Λ are related by the fact that both have the same spectrum of eigenvalues. The two matrices have different sets of eigenvectors. The orthonormal eigenvectors of Λ have a very simple structure: the ith eigenvector has zeros everywhere except for the ith element, which is equal to 1.

We have seen that the eigenvalues of a hermitian matrix are all real. Since an eigenvalue of an operator corresponding to a dynamical variable represents a possible result of the measurement of that variable, the eigenvalues of such an operator must be real. Therefore, we require that an operator corresponding to a dynamical variable be hermitian. This guarantees that the eigenvalues are real.

The matrix S that diagonalizes the matrix A in Example 1 is constructed from the normalized eigenvectors given in Eq. (9.5),

$$S = \frac{1}{\sqrt{2}} \begin{pmatrix} 0 & 1 & 1 \\ 0 & -1 & 1 \\ \sqrt{2} & 0 & 0 \end{pmatrix}.$$

The diagonalization procedure then yields,

$$S^{\dagger}AS = \frac{1}{\sqrt{2}} \begin{pmatrix} 0 & 0 & \sqrt{2} \\ 1 & -1 & 0 \\ 1 & 1 & 0 \end{pmatrix} \begin{pmatrix} 3 & 1 & 0 \\ 1 & 3 & 0 \\ 0 & 0 & 1 \end{pmatrix} \frac{1}{\sqrt{2}} \begin{pmatrix} 0 & 1 & 1 \\ 0 & -1 & 1 \\ \sqrt{2} & 0 & 0 \end{pmatrix}$$

$$= \frac{1}{2} \begin{pmatrix} 0 & 0 & \sqrt{2} \\ 1 & -1 & 0 \\ 1 & 1 & 0 \end{pmatrix} \begin{pmatrix} 0 & 2 & 4 \\ 0 & -2 & 4 \\ \sqrt{2} & 0 & 0 \end{pmatrix} = \begin{pmatrix} 1 & 0 & 0 \\ 0 & 2 & 0 \\ 0 & 0 & 4 \end{pmatrix}.$$

We see that the resultant matrix is diagonal with the eigenvalues of A distributed along the diagonal as expected.

9.3.4   INVERSE TRANSFORMATIONS

In rectangular coordinates, a position vector $\mathbf{r}$ can be represented by the column vector

$$\mathbf{r} = \mathbf{i}x + \mathbf{j}y + \mathbf{k}z = \begin{pmatrix} x \\ y \\ z \end{pmatrix}.$$

The operator Ω that rotates $\mathbf{r}$ through an angle ϕ about the z-axis[5] can be represented by the matrix

$$\Omega = \begin{pmatrix} \cos\phi & -\sin\phi & 0 \\ \sin\phi & \cos\phi & 0 \\ 0 & 0 & 1 \end{pmatrix}. \tag{9.12}$$

That is, $\Omega\mathbf{r} = \mathbf{r}'$, where $\mathbf{r}'$ denotes the vector obtained by rotating $\mathbf{r}$.

A generalization of this idea to transformation of a vector in an n-dimensional space is straightforward and useful with extensive applications both in classical and quantum physics. Let X and Y denote n-dimensional vectors

$$X = \begin{pmatrix} x_1 \\ x_2 \\ \vdots \\ x_n \end{pmatrix} \qquad \text{and} \qquad Y = \begin{pmatrix} y_1 \\ y_2 \\ \vdots \\ y_n \end{pmatrix},$$

---

[5]For definiteness, we take the direction of rotation about a coordinate axis to correspond to the direction such that a right-handed screw advances along the positive axis. Inverse rotations have the opposite sense.

related by a transformation matrix A such that

$$Y = AX,$$

where the elements (components) of Y are obtained from those of X by the rule of Eq. (9.3),

$$y_i = (AX)_i = \sum_{k=1}^{n} a_{ik} x_k.$$

In analogy with three-dimensional rotations, we expect that by a reverse operation, the vector Y can be rotated back so as to coincide with the original vector X. That is, there exists a transformation B such that

$$X = BY,$$

with

$$x_j = (BY)_j = \sum_{i=1}^{n} b_{ji} y_i = \sum_{i=1}^{n} b_{ji} \sum_{k=1}^{n} a_{ik} x_k$$

$$= \sum_{k=1}^{n} \left(\sum_{i=1}^{n} b_{ji} a_{ik} \right) x_k = \sum_{k=1}^{n} (BA)_{jk} x_k.$$

Clearly, if BY is to coincide with X, we must have

$$(BA)_{jk} = \delta_{jk}$$

or

$$BA = 1.$$

With this property, B is called the *inverse* of A and is denoted by $B \equiv A^{-1}$. That is,

$$AA^{-1} = A^{-1}A = 1.$$

Example 3.
Construct the inverse Ω^{-1} of the matrix Ω in Eq. (9.12).

Solution.
Since $\mathbf{r}'$ is obtained by rotating $\mathbf{r}$ through an angle ϕ about the z-axis, we expect that by rotating $\mathbf{r}'$ through $-\phi$ about this same axis, the original vector $\mathbf{r}$ is recovered. Hence, Ω^{-1} is obtained by replacing ϕ by $-\phi$ in Eq. (9.12). That is,

$$\Omega^{-1} = \begin{pmatrix} \cos\phi & \sin\phi & 0 \\ -\sin\phi & \cos\phi & 0 \\ 0 & 0 & 1 \end{pmatrix}. \tag{9.13}$$

Using direct matrix multiplication, it is easily verified that

$$\Omega^{-1}\Omega = \Omega\Omega^{-1} = 1.$$

Let C be a matrix that effects the rotation of a vector X about some axis. Let D be the matrix that performs a second rotation about a different axis resulting in the final transformed vector

$$Y = DCX. \tag{9.14}$$

Now multiply both sides of Eq. (9.14) with the matrix product $C^{-1}D^{-1}$,

$$C^{-1}D^{-1}Y = C^{-1}D^{-1}DCX = X. \tag{9.15}$$

According to our definition of an inverse, from Eq. (9.14)

$$(DC)^{-1}Y = (DC)^{-1}(DC)X = X. \tag{9.16}$$

On comparing Eqs. (9.15) and (9.16), we obtain the general rule for the inverse of a product of two matrices,

$$(DC)^{-1} = C^{-1}D^{-1}. \tag{9.17}$$

That is, the inverse of a product of two matrices is equal to the product of the individual inverses taken in reverse order. By successive applications, this rule can be extended to the product of any number of matrices. For example, for three matrices

$$[ABC]^{-1} = [(AB)C]^{-1} = C^{-1}(AB)^{-1} = C^{-1}B^{-1}A^{-1}. \tag{9.18}$$

9.3.5   A UNITARY MATRIX

A matrix U with the property

$$U^{\dagger}U = UU^{\dagger} = 1$$

is said to be *unitary*. Clearly, for a unitary matrix

$$U^{-1} = U^{\dagger} \qquad \text{(for } U \text{ unitary)}.$$

9.3.6   DEGENERACY

For a given hermitian matrix, it may happen that not all of the eigenvalues are different, that is, we may have $\lambda_i = \lambda_j$ for $i \neq j$. Suppose for an $n \times n$ hermitian matrix A, m of the n eigenvectors X_j have the same eigenvalue λ_k. That is,

$$\lambda_{k+i} = \lambda_k \qquad \text{for} \quad 0 \leq i \leq m - 1.$$

The eigenvectors, X_j with $k \leq j \leq k+m-1$, are said to be *degenerate*. There is an m-fold degeneracy among the eigenvectors corresponding to eigenvalues with the same value λ_k. In this case, the assumption that led us to conclude that eigenvectors corresponding to distinct eigenvalues are orthogonal is violated and we may have for these eigenvectors

$$X_i^{\dagger}X_j \neq 0 \qquad \text{for} \quad i \neq j.$$

Each of these vectors is orthogonal to any eigenvector corresponding to an eigenvalue different from λ_k,

$$X_i^\dagger X_j = X_j^\dagger X_i = 0 \qquad \text{for} \quad k \leq j \leq k+m-1$$
$$\text{and} \quad \left[(i \leq k-1) \quad \text{or} \quad (k+m \leq i \leq n)\right].$$

The eigenvectors X_j corresponding to $\lambda_j = \lambda_k$ are not orthogonal among themselves, but each of these eigenvectors is orthogonal to all eigenvectors X_i with i outside the range $k \leq i \leq k+m-1$. Therefore, these vectors do form a complete subset and from them we can construct m mutually orthogonal vectors that are automatically orthogonal to all the other $n-m$ eigenvectors of A. To illustrate how this can be done, let us start with one of the eigenvectors, say X_k, as the first vector in a new set X_j',

$$X_k' = X_k.$$

From this, we construct a normalized vector,

$$Y_k = \frac{X_k'}{\sqrt{X_k'^\dagger X_k'}} = \frac{X_k}{\sqrt{X_k^\dagger X_k}},$$

with $Y_k^\dagger Y_k = 1$. For the second vector in our new set, we take

$$X_{k+1}' = X_{k+1} - a_{10} Y_k,$$

and require that

$$Y_k^\dagger X_{k+1}' = Y_k^\dagger X_{k+1} - a_{10} = 0,$$

from which we get $a_{10} = Y_k^\dagger X_{k+1}$. The corresponding normalized vector is

$$Y_{k+1} = \frac{X_{k+1}'}{\sqrt{X_{k+1}'^\dagger X_{k+1}'}}.$$

For the third vector in the new set, we take

$$X_{k+2}' = X_{k+2} - a_{21} Y_{k+1} - a_{20} Y_k,$$

requiring

$$Y_k^\dagger X_{k+2}' = Y_k^\dagger X_{k+2} - a_{20} = 0,$$
$$\text{and} \qquad Y_{k+1}^\dagger X_{k+2}' = Y_{k+1}^\dagger X_{k+2} - a_{21} = 0,$$

yielding $a_{20} = Y_k^\dagger X_{k+2}$ and $a_{21} = Y_{k+1}^\dagger X_{k+2}$. The third normalized vector is then,

$$Y_{k+2} = \frac{X_{k+2}'}{\sqrt{X_{k+2}'^\dagger X_{k+2}'}}.$$

The general pattern is now clear. The new eigenvector X'_{k+i} in the subset corresponding to the m eigenvalues with the same value λ_k is

$$X'_{k+i} = X_{k+i} - \sum_{j=0}^{i-1} a_{ij} Y_{k+j},$$

with coefficients given by

$$a_{ij} = Y^\dagger_{k+j} X_{k+i} \qquad \text{for} \quad 0 \le j \le i-1.$$

The corresponding member of the orthonormal set is

$$Y_{k+i} = \frac{X'_{k+i}}{\sqrt{X'^\dagger_{k+i} X'_{k+i}}}.$$

The vectors Y_i constructed in this way have the properties,

$$Y^\dagger_i Y_j = \delta_{ij} \qquad \text{for} \quad k \le j \le k+m-1, \text{ and}$$
$$Y^\dagger_i X_j = 0 \qquad \text{for} \quad (j < k) \quad \text{or} \quad (k+m \le j \le n).$$

This method of constructing m orthonormal vectors Y_i from a set of m independent vectors X_j is known as *Schmidt's orthogonalization method*.[6]

Example 4.
From the three vectors,

$$X_1 = \begin{pmatrix} 5 \\ -1 \\ 3 \end{pmatrix} \qquad X_2 = \begin{pmatrix} 4 \\ 2 \\ 7 \end{pmatrix} \qquad X_3 = \begin{pmatrix} -8 \\ 3 \\ 4 \end{pmatrix}$$

construct three orthonormal vectors Y_j using the Schmidt orthogonalization procedure.

Solution.
Choosing $X'_1 = X_1$, we get

$$Y_1 = \frac{X'_1}{\sqrt{X'^\dagger_1 X'_1}} = \sqrt{\frac{1}{35}} \begin{pmatrix} 5 \\ -1 \\ 3 \end{pmatrix}.$$

Next, we write $X'_2 = X_2 - (Y^\dagger_1 X_2)Y_1$, from which we obtain $X'^\dagger_2 X'_2 = 894/35$ and

$$Y_2 = \frac{X'_2}{\sqrt{X'^\dagger_2 X'_2}} = \sqrt{\frac{1}{31,290}} \begin{pmatrix} -55 \\ 109 \\ 128 \end{pmatrix}.$$

---

[6]See Margenau and Murphy, *op.cit.*, p. 312.

For the third vector, $X_3' = X_3 - (Y_2^\dagger X_3)Y_2 - (Y_1^\dagger X_3)Y_1$. Calculating the coefficients of Y_1 and Y_2, we find $X_3'^\dagger X_3' = 8281/894$ and the normalized vector,

$$Y_3 = \frac{X_3'}{\sqrt{X_3'^\dagger X_3'}} = \sqrt{\frac{1}{7403214}} \begin{pmatrix} -1183 \\ -2093 \\ 1274 \end{pmatrix}.$$

9.4 Coupled Harmonic Oscillators

Now we turn to the matrix formulation of an eigenvalue problem that arises in classical mechanics. Two identical particles of mass m are bound by identical elastic springs as illustrated in Fig. 9.1. According to Newton's second law, the equations of motion for these particles are

$$m\frac{d^2x}{dt^2} + k(2x - y) = 0$$

$$m\frac{d^2y}{dt^2} + k(2y - x) = 0, \qquad (9.19)$$

where x and y are the displacements of the masses from their equilibrium positions as indicated in Fig. 9.1.

Because the springs are elastic, we expect the motion to be oscillatory. We are particularly interested in those modes of oscillation for which the two masses have the same frequency. These *normal* modes will represent a complete set. Hence, any arbitrary state of motion of the oscillating system can be expressed as a superposition of these two. Thus, we assume solutions to Eq. (9.19) of the form,

$$x(t) = ae^{i\omega t} \qquad \text{and} \qquad y(t) = be^{i\omega t},$$

where the amplitudes a and b are complex and ω is a real, positive constant. Substituting these expressions into Eq. (9.19), we obtain the coupled algebraic equations

$$-\omega^2 a + \omega_0^2(2a - b) = 0$$
$$-\omega^2 b + \omega_0^2(2b - a) = 0, \qquad (9.20)$$

where we have introduced the parameter $\omega_0 \equiv \sqrt{k/m}$.

Writing Eqs. (9.20) in matrix form, we have

$$\begin{pmatrix} 2\omega_0^2 - \omega^2 & -\omega_0^2 \\ -\omega_0^2 & 2\omega_0^2 - \omega^2 \end{pmatrix} \begin{pmatrix} a \\ b \end{pmatrix} = 0. \qquad (9.21)$$

To obtain nontrivial solutions for the amplitudes a and b, we require that the determinant of the 2×2 matrix in Eq. (9.21) vanish. That is,

$$(2\omega_0^2 - \omega^2)^2 - \omega_0^4 = 0.$$

The four roots of this equation are easily seen to be $\omega = \pm \omega_0$ and $\omega = \pm\sqrt{3}\,\omega_0$. From these results, we obtain two sets of solutions to Eq. (9.19), namely:

$$x_1(t) = a_1^{(+)} e^{i\omega_0 t} + a_1^{(-)} e^{-i\omega_0 t} \qquad\qquad y_1(t) = b_1^{(+)} e^{i\omega_0 t} + b_1^{(-)} e^{-i\omega_0 t}$$

and

$$x_2(t) = a_2^{(+)} e^{i\sqrt{3}\omega_0 t} + a_2^{(-)} e^{-i\sqrt{3}\omega_0 t} \qquad y_2(t) = b_2^{(+)} e^{i\sqrt{3}\omega_0 t} + b_2^{(-)} e^{-i\sqrt{3}\omega_0 t}.$$

Substituting each of the eigenfrequencies for ω in Eq. (9.21) to obtain the corresponding eigenvectors, we get

$$b_1^{(\pm)} = a_1^{(\pm)} \qquad\text{and}\qquad b_2^{(\pm)} = -a_2^{(\pm)}.$$

Finally, we obtain the general solution to Eq. (9.19),

$$x(t) = a_1^{(+)} e^{i\omega_0 t} + a_1^{(-)} e^{-i\omega_0 t} + a_2^{(+)} e^{i\sqrt{3}\omega_0 t} + a_2^{(-)} e^{-i\sqrt{3}\omega_0 t}$$
$$y(t) = a_1^{(+)} e^{i\omega_0 t} + a_1^{(-)} e^{-i\omega_0 t} - a_2^{(+)} e^{i\sqrt{3}\omega_0 t} - a_2^{(-)} e^{-i\sqrt{3}\omega_0 t}. \qquad (9.22)$$

There are four arbitrary constants of integration in this solution, as expected since we have two second-order differential equations in Eq. (9.19).

Suppose we hold the mass on the left in Fig. 9.1 at its equilibrium position and pull the mass on the right a distance R to the right of its equilibrium position. Now, we release both masses simultaneously from rest. Imposing these initial conditions at time $t = 0$, we get from Eq. (9.22) four equations for the four amplitudes,

$$x(0) = a_1^{(+)} + a_1^{(-)} + a_2^{(+)} + a_2^{(-)} = 0$$
$$y(0) = a_1^{(+)} + a_1^{(-)} - a_2^{(+)} - a_2^{(-)} = R$$
$$\left.\frac{dx(t)}{dt}\right|_{t=0} = i\omega_0\big[a_1^{(+)} - a_1^{(-)} + \sqrt{3}a_2^{(+)} - \sqrt{3}a_2^{(-)}\big] = 0$$
$$\left.\frac{dy(t)}{dt}\right|_{t=0} = i\omega_0\big[a_1^{(+)} - a_1^{(-)} - \sqrt{3}a_2^{(+)} + \sqrt{3}a_2^{(-)}\big] = 0.$$

A straightforward solution of these equations yields

$$a_1^{(\pm)} = -a_2^{(\pm)} = \tfrac{1}{4}R.$$

With these values for the amplitudes in the expression for $x(t)$, we get

$$x(t) = \tfrac{1}{4}R\big(e^{i\omega_0 t} + e^{-i\omega_0 t}\big) - \tfrac{1}{4}R\big(e^{i\sqrt{3}\omega_0 t} + e^{-i\sqrt{3}\omega_0 t}\big)$$
$$= \tfrac{1}{2}R\big(\cos\omega_0 t - \cos\sqrt{3}\omega_0 t\big).$$

Similarly, we obtain for $y(t)$,

$$y(t) = \tfrac{1}{2}R\big(\cos\omega_0 t + \cos\sqrt{3}\omega_0 t\big).$$

It is easily checked that these expressions satisfy the initial conditions at $t = 0$. We see that on setting the system in motion in this way, the oscillations of the two masses are represented by linear combinations of the two normal modes.

Now let's see what happens if we start the motion with a different set of initial conditions. Suppose we pull each of the masses in Fig. 9.1 a distance R to the

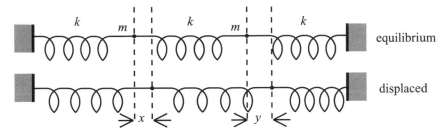

FIGURE 9.1. Identical coupled harmonic oscillators.

right of its equilibrium position and release both masses simultaneously from rest. Following the same procedure as before leads to four new equations for the amplitudes,

$$x(0) = a_1^{(+)} + a_1^{(-)} + a_2^{(+)} + a_2^{(-)} = R$$

$$y(0) = a_1^{(+)} + a_1^{(-)} - a_2^{(+)} - a_2^{(-)} = R$$

$$\left.\frac{dx(t)}{dt}\right|_{t=0} = i\omega_0\left[a_1^{(+)} - a_1^{(-)} + \sqrt{3}a_2^{(+)} - \sqrt{3}a_2^{(-)}\right] = 0$$

$$\left.\frac{dy(t)}{dt}\right|_{t=0} = i\omega_0\left[a_1^{(+)} - a_1^{(-)} - \sqrt{3}a_2^{(+)} + \sqrt{3}a_2^{(-)}\right] = 0.$$

From these equations, we find that

$$a_1^{(\pm)} = \tfrac{1}{2}R \qquad \text{and} \qquad a_2^{(\pm)} = 0.$$

Substituting these results into the general expressions for $x(t)$ and $y(t)$ in Eq. (9.22), we obtain the solution

$$x(t) = R \cos \omega_0 t$$

$$y(t) = R \cos \omega_0 t.$$

We see that by setting the masses in motion as described in this case, the two masses oscillate in phase (i.e., in step together) with a common frequency $\omega = \omega_0$. They behave like a single oscillator with a frequency equal to that of one of the normal modes. Starting the oscillations in this particular way excites one of the normal modes. It is left as an exercise to find the initial conditions required to excite the other normal mode.

9.5   A Rotating Rigid Body

The rectangular components of the angular momentum **L** of a rigid body rotating about a fixed axis are given by

$$L_i = \sum_{j=1}^{3} I_{ij}\omega_j, \tag{9.23}$$

where ω_j is the jth rectangular component of the angular velocity ω about the rotation axis. The nine-component quantity I with elements I_{ij} is called the *moment of inertia tensor*. These elements depend on the distribution of mass in the body and on the choice of the origin of the coordinate system relative to the body. With the position vector $\mathbf{r}$ written as $\mathbf{r} = \mathbf{i}\,x_1 + \mathbf{j}\,x_2 + \mathbf{k}\,x_3$, the elements I_{ij} are calculated in rectangular coordinates according to,

$$I_{ij} = \int \rho(\mathbf{r})\Big[\delta_{ij}\sum_{k=1}^{3} x_k^2 - x_i x_j\Big]d^3r, \tag{9.24}$$

where $\rho(\mathbf{r})$ is the mass density of the body. It is easily seen that these elements have the property $I_{ij} = I_{ji}$.

The relation between angular momentum and angular velocity given in Eq. (9.23) can be expressed as a matrix equation,

$$\begin{pmatrix} L_1 \\ L_2 \\ L_3 \end{pmatrix} = \begin{pmatrix} I_{11} & I_{12} & I_{13} \\ I_{21} & I_{22} & I_{23} \\ I_{31} & I_{32} & I_{33} \end{pmatrix} \begin{pmatrix} \omega_1 \\ \omega_2 \\ \omega_3 \end{pmatrix}. \tag{9.25}$$

In general, the off-diagonal elements I_{ij} are not equal to zero. Therefore, for rotations about an arbitrary axis, the angular momentum may not be along the same direction as the angular velocity. However, for some particular directions of the angular velocity it can happen that the angular momentum $\mathbf{L}$ is in the same direction as the angular velocity ω. For such an axis of rotation, we have

$$\mathbf{L} = I\omega, \tag{9.26}$$

where I is a scalar. An axis of rotation with this feature is called a *principal axis* of the body. It is important to remember that this special relation between $\mathbf{L}$ and ω only holds for certain directions of ω relative to the body.

For rotation about a principal axis, we combine Eqs. (9.25) and (9.26) to get

$$\begin{pmatrix} L_1 \\ L_2 \\ L_3 \end{pmatrix} = \begin{pmatrix} I_{11} & I_{12} & I_{13} \\ I_{21} & I_{22} & I_{23} \\ I_{31} & I_{32} & I_{33} \end{pmatrix} \begin{pmatrix} \omega_1 \\ \omega_2 \\ \omega_3 \end{pmatrix} = I \begin{pmatrix} \omega_1 \\ \omega_2 \\ \omega_3 \end{pmatrix}. \tag{9.27}$$

This is an eigenvalue problem. Find the values of I (eigenvalues) and corresponding directions of ω (eigenvectors) for which this relation holds. This eigenvalue problem can also be written as

$$\begin{pmatrix} I_{11} - I & I_{12} & I_{13} \\ I_{21} & I_{22} - I & I_{23} \\ I_{31} & I_{32} & I_{33} - I \end{pmatrix} \begin{pmatrix} \omega_1 \\ \omega_2 \\ \omega_3 \end{pmatrix} = 0. \tag{9.28}$$

To obtain the eigenvalues, we set the determinant of the 3×3 matrix in Eq. (9.28) equal to zero and solve for the three roots I.

Example 5.

A cube of uniform density with length s along a side has a total mass M. Find the principal moments of inertia for the cube and the directions of the corresponding axes of rotation.

Solution.
Choose the coordinate system so that the cube lies entirely in the first octant with the origin at one corner of the cube. Using Eq. (9.24) to calculate the elements of the moment of inertia tensor, we get

$$\mathsf{I} = \frac{Ms^2}{12} \begin{pmatrix} 8 & -3 & -3 \\ -3 & 8 & -3 \\ -3 & -3 & 8 \end{pmatrix}. \tag{9.29}$$

To find the eigenvalues, we set the determinant

$$\begin{vmatrix} 8-\lambda & -3 & -3 \\ -3 & 8-\lambda & -3 \\ -3 & -3 & 8-\lambda \end{vmatrix} = 242 - 165\lambda + 24\lambda^2 - \lambda^3 = 0. \tag{9.30}$$

Using the method in Appendix D, we find the roots of this equation to be $\lambda = 2, 11, 11$. From Eqs. (9.29) and (9.30), we see that the moments of inertia about the principal axes of the cube are

$$I = \frac{Ms^2}{12}\lambda = \frac{1}{6}Ms^2; \ \frac{11}{12}Ms^2; \ \frac{11}{12}Ms^2.$$

The directions of the principal axes of rotation are given by the corresponding eigenvectors. For example, the direction of the principal axis corresponding to $I = Ms^2/6$ (i.e., $\lambda = 2$) is obtained from

$$\begin{pmatrix} 6 & -3 & -3 \\ -3 & 6 & -3 \\ -3 & -3 & 6 \end{pmatrix} \begin{pmatrix} a \\ b \\ c \end{pmatrix} = \begin{pmatrix} 6a - 3b - 3c \\ -3a + 6b - 3c \\ -3a - 3b + 6c \end{pmatrix} = \begin{pmatrix} 0 \\ 0 \\ 0 \end{pmatrix},$$

from which we find that $a = b = c$. Thus, for rotations about this principal axis, each of the three components of the angular velocity vector ω is equal to each of the other two. That is, the principal axis is along the diagonal of the cube. In this case, we can write Eq. (9.27) as

$$\begin{pmatrix} L_1 \\ L_2 \\ L_3 \end{pmatrix} = \frac{Ms^2\omega}{6\sqrt{3}} \begin{pmatrix} 1 \\ 1 \\ 1 \end{pmatrix},$$

where ω is the magnitude of the angular velocity.

The other two moments of inertia are equal. Therefore, any two perpendicular vectors that lie in a plane perpendicular to the diagonal (i.e., the first principal axis) will serve as mutually perpendicular eigenvectors corresponding to these two moments of inertia (eigenvalues). From this, we deduce that the first principal axis is a symmetry axis for the cube.

9.6 Exercises

1. For the two matrices A and B given below, construct the four products AB, BA, $A^\dagger B$, and $B^\dagger A$.

$$A = \begin{pmatrix} 1 & i & 0 \\ i & 2 & 0 \\ 0 & 0 & 3 \end{pmatrix} \quad \text{and} \quad B = \begin{pmatrix} 4 & 7i & 0 \\ 2i & 7 & 0 \\ 0 & 0 & 2 \end{pmatrix}.$$

Examine these products for similarities and differences among them.

2. Find the eigenvalues and the corresponding *normalized* eigenvectors for each of the following matrices. In each case, construct the matrix S that diagonalizes A according to

$$S^\dagger A S = \Lambda,$$

where Λ is a diagonal matrix whose elements are the eigenvalues of A.

   a. $A = \begin{pmatrix} 1 & 2 \\ 2 & -2 \end{pmatrix}$ b. $A = \begin{pmatrix} -3 & 6 \\ 6 & 2 \end{pmatrix}$ c. $A = \begin{pmatrix} 1 & 2 \\ 2 & 4 \end{pmatrix}$

   d. $A = \begin{pmatrix} 1 & 0 & -2 \\ 0 & 3 & 0 \\ -2 & 0 & 4 \end{pmatrix}$ e. $A = \begin{pmatrix} -2 & 1 & 0 \\ 1 & 1 & -3 \\ 0 & -3 & -2 \end{pmatrix}$

   f. $A = \begin{pmatrix} -3 & 0 & 0 \\ 0 & 2 & 2 \\ 0 & 2 & -1 \end{pmatrix}$ g. $A = \begin{pmatrix} 5 & 0 & 0 \\ 0 & -6 & 3 \\ 0 & 3 & 2 \end{pmatrix}$

3. Obtain the other two values for x in the equation in Example 9.2. Verify that each of the three values obtained for x is a root of this equation.

4. Use the method described in Appendix D to find the values of x that satisfy the cubic equation

$$x^3 + 9x^2 + 15x + 7 = 0.$$

Note that this equation has a double root (i.e., two of the values of x are the same).

5. We can see by inspection that the three roots of the equation

$$x^3 - x^2 + x - 1 = 0$$

are $x = 1$ and $x = \pm i$. Use the method outlined in Appendix D to obtain the roots of this equation and show explicitly that these roots are equivalent to the ones obtained by inspection.
   HINT: Verify $(1 + \sqrt{3})^3 = 10 + 6\sqrt{3}$.

6. Find the roots of the equation

$$x^3 + 6x^2 + 24x + 44 = 0.$$

By direct substitution, verify that your values for x satisfy the equation.

7. Find the eigenvalues and corresponding *normalized* eigenvectors for each of the matrices

$$\begin{pmatrix} 0 & i & 0 \\ -i & 0 & 0 \\ 0 & 0 & 0 \end{pmatrix} \qquad \begin{pmatrix} 2 & i & 0 \\ -i & 2 & 0 \\ 0 & 0 & -3 \end{pmatrix} \qquad \begin{pmatrix} 1 & 0 & i \\ 0 & 1 & 0 \\ -i & 0 & 1 \end{pmatrix}.$$

Write down explicitly the matrix S that diagonalizes each of these matrices A according to

$$S^\dagger A S = \Lambda,$$

where Λ is a matrix whose diagonal elements are the eigenvalues of A and whose off-diagonal elements are all equal to zero. Show explicitly that for each matrix A your corresponding S diagonalizes A in this sense.

8. One of the eigenvalues of the matrix

$$\begin{pmatrix} 3 & -2 & -1 \\ -2 & 3 & 1 \\ -1 & 1 & 2 \end{pmatrix}$$

is $+1$. Find the other two eigenvalues.

9. The matrix U is unitary. Show that the matrix T, defined by

$$T = U - 1,$$

has the property

$$T + T^\dagger = -T^\dagger T = -TT^\dagger.$$

10. The matrices A and B are both unitary. Show that the product AB is unitary.

11. Two matrices C and D are hermitian. Under what circumstances is the product CD hermitian? Explain clearly.

12. For the matrix operator Ω of Eq. (9.12), we have the eigenvalue equation

$$\Omega \mathbf{q} = \lambda \mathbf{q},$$

where $\mathbf{q}$ is the eigenvector for the eigenvalue λ. Find the eigenvalues and the corresponding eigenvectors. By matrix multiplication, verify that with Ω^{-1} given in Eq. (9.13), $\Omega\Omega^{-1} = \Omega^{-1}\Omega = 1$.

13. The matrix representation of an operator A corresponding to the rotation of a vector $\mathbf{r}$ through an angle α about the z-axis is

$$A(\alpha) = \begin{pmatrix} \cos\alpha & -\sin\alpha & 0 \\ \sin\alpha & \cos\alpha & 0 \\ 0 & 0 & 1 \end{pmatrix}.$$

Thus, $\mathbf{r}' = A\mathbf{r}$. Similarly, rotation of a vector $\mathbf{r}'$ through an angle β about the x-axis is achieved with the matrix operator

$$B(\beta) = \begin{pmatrix} 1 & 0 & 0 \\ 0 & \cos\beta & -\sin\beta \\ 0 & \sin\beta & \cos\beta \end{pmatrix},$$

with $\mathbf{r}'' = B\mathbf{r}'$. For successive rotations, first about the z-axis, then about the x-axis, we have $\mathbf{r}'' = B\mathbf{r}' = BA\mathbf{r}$. By matrix multiplication, construct the single matrix $C = BA$, such that

$$\mathbf{r}'' = C\mathbf{r}.$$

Verify that $A^{-1}(\alpha) = A(-\alpha)$ and $B^{-1}(\beta) = B(-\beta)$ and obtain the explicit construction of C^{-1}. Verify by direct matrix multiplication that

$$C^{-1}C = CC^{-1} = 1.$$

14. Use the rule for obtaining the inverse of a product of two matrices in Eq. (9.17) to show that for two $n \times n$ matrices A and B,

$$\frac{1}{A - B} = \frac{1}{A} + \frac{1}{A}B\frac{1}{A} + \frac{1}{A}B\frac{1}{A}B\frac{1}{A} + \dots,$$

assuming $A^{-1} \equiv 1/A$ exists.

15. Energy eigenfunctions for the electron in a hydrogen atom can be written in the form

$$u_{nlm_l}(\mathbf{r})\chi_s^{m_s},$$

where $u_{nlm_l}(\mathbf{r})$ gives the spatial part of the eigenfunction (from Schrödinger's equation) and $\chi_s^{m_s}$ represents the intrinsic angular momentum (spin) of the electron. The function $\chi_s^{m_s}$ satisfies the eigenvalue equations

$$S^2\chi_s^{m_s} = s(s + 1)\hbar^2\chi_s^{m_s}$$

and

$$S_z\chi_s^{m_s} = m_s\hbar\chi_s^{m_s}$$

where S^2 and S_z are operators corresponding, respectively, to the square of the spin and the projection of the spin onto the z-axis. As is the case here, angular momenta in quantum mechanics are expressed in units of $\hbar$. For all electrons, $s = \frac{1}{2}$, hence m_s has only two possible values, $m_s = \pm\frac{1}{2}$. It is convenient to treat the spin in terms of matrices. For example,

$$S^2 = \frac{3}{4}\hbar^2\begin{pmatrix} 1 & 0 \\ 0 & 1 \end{pmatrix} \qquad \text{and} \qquad S_z = \frac{1}{2}\hbar\begin{pmatrix} 1 & 0 \\ 0 & -1 \end{pmatrix}.$$

For a general direction of the spin, we can write the spin vector operator, $\mathbf{S}$ as

$$\mathbf{S} = \frac{1}{2}\boldsymbol{\sigma}.$$

In the so-called Pauli representation above, the rectangular components of σ are

$$\sigma_x = \begin{pmatrix} 0 & 1 \\ 1 & 0 \end{pmatrix} \qquad \sigma_y = \begin{pmatrix} 0 & -i \\ i & 0 \end{pmatrix} \qquad \sigma_z = \begin{pmatrix} 1 & 0 \\ 0 & -1 \end{pmatrix}.$$

These are the Pauli spin matrices. Find the eigenvalues and corresponding (normalized) eigenvectors for each of the matrices σ_x, σ_y, and σ_z. For each of these, construct a matrix U that diagonalizes σ_i according to

$$U^\dagger \sigma_i U = \rho_i$$

where ρ_i is a 2×2 matrix with nonzero elements only on the diagonal. Is U the same for all σ_i? What implication does this have for simultaneously diagonalizing σ_x, σ_y, and σ_z?

16. Let **n** be an arbitrary unit vector,

$$\mathbf{n} = \mathbf{i} n_x + \mathbf{j} n_y + \mathbf{k} n_z.$$

Use the representation of the Pauli matrices given in Exercise 9.15 to obtain the eigenvalues and corresponding normalized eigenvectors for the 2×2 matrix $\sigma \cdot \mathbf{n}$. Construct the matrix R such that

$$R^\dagger \sigma \cdot \mathbf{n} R = \Lambda.$$

where Λ is the 2×2 diagonal matrix whose nonzero elements are the eigenvalues of $\sigma \cdot \mathbf{n}$. Verify that R diagonalizes $\sigma \cdot \mathbf{n}$ according to this rule.

17. The three 2×2 Pauli spin matrices defined in Exercise 9.15 have the properties[7]

- $\sigma_i \sigma_j + \sigma_j \sigma_i = 2\delta_{ij}$,
- $\sigma_2 \sigma_3 = i\sigma_1$ $\qquad$ $\sigma_3 \sigma_1 = i\sigma_2$ $\qquad$ $\sigma_1 \sigma_2 = i\sigma_3$.

Let **a** and **b** be any two arbitrary vectors that commute with σ, but not necessarily with each other. Using the properties given above, show that

$$\sigma \cdot \mathbf{a}\, \sigma \cdot \mathbf{b} = \mathbf{a} \cdot \mathbf{b} + i\sigma \cdot (\mathbf{a} \times \mathbf{b}).$$

18. Verify that each of the Pauli matrices in Exercise 9.15 is both hermitian and unitary.

19. Verify that the three vectors Y_1, Y_2, and Y_3 obtained in Example 9.4 are orthonormal.

20. From the four vectors,

$$X_1 = \begin{pmatrix} 1 \\ 1 \\ 1 \\ 1 \end{pmatrix} \qquad X_2 = \begin{pmatrix} 1 \\ -2 \\ 1 \\ 1 \end{pmatrix} \qquad X_3 = \begin{pmatrix} 2 \\ 1 \\ 1 \\ 1 \end{pmatrix} \qquad X_4 = \begin{pmatrix} 2 \\ 1 \\ 2 \\ -1 \end{pmatrix}$$

construct four orthonormal vectors Y_j using the Schmidt orthogonalization procedure. Check your vectors Y_j for normalization to unity and orthogonality.

21. If any one of three arbitrary vectors U_1, U_2, U_3 can be be written as a linear combination of the other two, the three vectors are not independent. Show that the vectors

$$U_1 = \begin{pmatrix} 3 \\ -1 \\ 2 \end{pmatrix} \qquad U_2 = \begin{pmatrix} -1 \\ 2 \\ 1 \end{pmatrix} \qquad U_3 = \begin{pmatrix} 1 \\ 2 \\ 3 \end{pmatrix}$$

---

[7] Here we replace the indices x, y, z in Exercise 9.15 with 1, 2, 3, respectively.

are not independent by finding values for a and b such that

$$a\,U_1 + b\,U_2 = U_3.$$

22. Show that the three vectors

$$X_1 = \begin{pmatrix} 3 \\ -1 \\ 2 \end{pmatrix} \qquad X_2 = \begin{pmatrix} 1 \\ -1 \\ 2 \end{pmatrix} \qquad X_3 = \begin{pmatrix} 1 \\ 2 \\ 3 \end{pmatrix}$$

are independent. Starting with X_1, use the Schmidt procedure to construct from these vectors a set of three orthonormal vectors. Then, starting with X_3, construct another set of orthonormal vectors from X_1, X_2, and X_3. Check the vectors in each set for normalization and orthogonality.

23. Show that the three vectors

$$X_1 = \begin{pmatrix} 1 \\ i \\ -3i \end{pmatrix} \qquad X_2 = \begin{pmatrix} 2i \\ -i \\ 3 \end{pmatrix} \qquad X_3 = \begin{pmatrix} i \\ 2i \\ 2 \end{pmatrix}$$

are independent. Verify that none of the vectors is orthogonal to either of the other two. Construct an orthonormal set of vectors from X_1, X_2, and X_3.

24. There are two normal modes for the system of two coupled harmonic oscillators described in Section 9.4. We considered the initial conditions required to excite the normal mode with eigenfrequency ω_0. Find the initial conditions for exciting the other normal mode with eigenfrequency $\sqrt{3}\,\omega_0$.

25. Suppose the middle spring in Fig. 9.1 has a force constant k_1 different from the force constant k for the other two springs. Show that the corresponding equations of motion are

$$m\frac{d^2x}{dt^2} + (k + k_1)x - k_1 y = 0$$

$$m\frac{d^2y}{dt^2} + (k + k_1)y - k_1 x = 0.$$

Assume harmonic solutions

$$x(t) = ae^{i\omega t} \qquad \text{and} \qquad y(t) = be^{i\omega t}.$$

Follow the procedure in Section 9.4 to show that the equations for the amplitudes a and b can be reduced to matrix form

$$\begin{pmatrix} \omega_0^2 + \omega_1^2 - \omega^2 & -\omega_1^2 \\ -\omega_1^2 & \omega_0^2 + \omega_1^2 - \omega^2 \end{pmatrix} \begin{pmatrix} a \\ b \end{pmatrix} = 0,$$

with the new parameter $\omega_1 \equiv \sqrt{k_1/m}$. Find the eigenvalues ω for this system and construct the general solution to the equation of motion for $x(t)$ and $y(t)$.

26. Three identical particles each of mass m are bound by four identical elastic springs as illustrated in Fig. 9.2. Write down the three equations of motion for this system analogous to Eq. (9.19). Follow the procedure in Section 9.4 to find the three eigenfrequencies for this three-particle system.

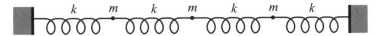

FIGURE 9.2. Three coupled harmonic oscillators for Exercise 9.26.

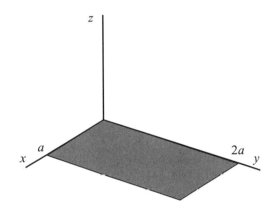

FIGURE 9.3. A thin rectangular sheet for Exercise 9.28.

27. Use Eq. (9.24) to calculate the matrix elements of the moment of inertia tensor given in Eq. (9.29).

28. A thin, flat rectangular sheet lies in the xy-plane as illustrated in Fig. 9.3. The sheet is of uniform density and is twice as long as it is wide. Use Eq. (9.24) to show that the moment of inertia tensor for this sheet is

$$
\mathsf{I} = \frac{Ma^2}{6} \begin{pmatrix} 8 & -3 & 0 \\ -3 & 2 & 0 \\ 0 & 0 & 10 \end{pmatrix},
$$

where M is the total mass of the sheet and a is the length of the shorter side. Find the eigenvalues (principal moments of inertia of the sheet) and corresponding eigenvectors (principal axes of rotation) for this matrix. Show that one of the principal axes is parallel to the z-axis. Show that the other two axes lie in the xy-plane with one making an angle of exactly $22\frac{1}{2}°$ with the x-axis. Show that the third principal axis is perpendicular to the first two.

HINT: Assume the sheet has a finite thickness τ with the bottom of the sheet lying in the xy-plane. To get the tensor elements I_{ij}, take the limit as $\tau \to 0$.

29. Three point masses m_1, m_2, and m_3 are located on the axes of a rectangular coordinate system according to

$$
\begin{aligned}
m_1 &\quad \text{at} \quad (a, 0, 0), \\
m_2 &\quad \text{at} \quad (0, b, 0), \\
m_3 &\quad \text{at} \quad (0, 0, c).
\end{aligned}
$$

The mass density of this system is given by

$$\rho(\mathbf{r}) = \sum_{l=1}^{3} m_l \delta^3(\mathbf{r} - \mathbf{r}_l).$$

Use Eq. (9.24) to calculate the elements I_{ij} of the moment of inertia tensor for this three-particle system. Give the moment of inertia of the system about each of these axes.

30. A rigid structure consists of three point masses. The rectangular coordinates of each of the masses are given in the table. Construct the moment of inertia matrix by following a procedure similar to that described in Exercise 9.29. Find the directions of the principal axes and the moment of inertia about each of these axes.

mass (g)	x (cm)	y (cm)	z (cm)
5	2	1	4
3	3	4	2
2	4	3	1

31. Find the eigenvalues of the matrix

$$\begin{pmatrix} 1 & -1 & -2 \\ -1 & 2 & -3 \\ -2 & -3 & 3 \end{pmatrix}.$$

Verify that these eigenvalues are real. Show by direct substitution that each eigenvalue satisfies the cubic equation from which it was obtained as a solution.

32. Find the eigenvalues and corresponding normalized eigenvectors of the matrix

$$\begin{pmatrix} 0 & -1 & 1 \\ -1 & 0 & -2 \\ 1 & -2 & 0 \end{pmatrix}.$$

33. We found the eigenvalues of the matrix in Example 9.5,

$$A = \begin{pmatrix} 8 & -3 & -3 \\ -3 & 8 & -3 \\ -3 & -3 & 8 \end{pmatrix},$$

to be $\lambda = 2, 11, 11$ implying a two-fold degeneracy among the eigenvectors. Verify that the vectors

$$X_1 = \begin{pmatrix} 1 \\ 1 \\ 1 \end{pmatrix} \qquad X_2 = \begin{pmatrix} 1 \\ 0 \\ -1 \end{pmatrix} \qquad X_3 = \begin{pmatrix} 4 \\ -5 \\ 1 \end{pmatrix}$$

are eigenvectors of A. Use the Schmidt orthogonalization method to construct a set of three orthonormal eigenvectors Y_j. Verify that your vectors Y_j are mutually orthogonal, are normalized to unity, and are eigenvectors of A.

34. Find the eigenvalues of the matrix

$$A = \begin{pmatrix} 1 & 1 & 1 \\ 1 & 1 & 1 \\ 1 & 1 & 1 \end{pmatrix},$$

and construct a corresponding set of orthonormal eigenvectors. Verify that your vectors are mutually orthogonal, are normalized to unity, and are eigenvectors of the matrix A.

35. Obtain the eigenvalues and a corresponding set of orthonormal eigenvectors for the matrix,

$$A = \begin{pmatrix} 2 & -1 & -1 \\ -1 & 2 & -1 \\ -1 & -1 & 2 \end{pmatrix}.$$

Verify that your vectors are mutually orthogonal, are normalized to unity, and are eigenvectors of the matrix A.

36. Obtain the eigenvalues of the hermitian matrix

$$A = \begin{pmatrix} 2 & -i & 1 \\ i & 2 & -i \\ 1 & i & 2 \end{pmatrix}.$$

Construct a set of orthonormal eigenvectors corresponding to these eigenvalues. Verify that your eigenvectors are mutually orthogonal and normalized to unity. Construct the matrix S that diagonalizes A according to

$$S^\dagger A S = \Lambda,$$

where Λ is a 3×3 matrix with all off-diagonal elements equal to zero.

10

Variational Principles

10.1 Fermat's Principle

Variational principles play an extremely important role in modern physics. To illustrate the idea, let us look at a simple example from optics involving the reflection and refraction of light at the boundary between two transparent media. What path does the light take in each case? For example, light is emitted from a source S, reflected from a boundary, and detected at a point F. Some possible paths along which the light might propagate to get from S to F are indicated by the dashed lines in Fig. 10.1. Which of these, if any, represents the *actual* path of the light? We seek an answer by invoking a variational principle known as *Fermat's principle*. One statement of Fermat's principle that will serve our purpose for the present

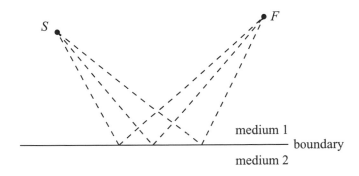

FIGURE 10.1. Path of light on reflection at a boundary.

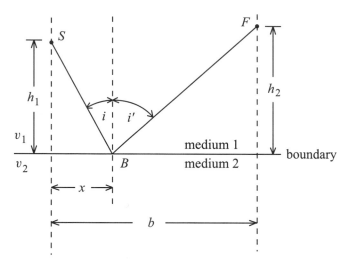

FIGURE 10.2. The law of reflection.

is that *the actual path taken by a beam of light between two points is the one that requires the shortest time for the light to travel from one point to the other*. In medium 1, the propagation velocity of light is v_1 and in medium 2 it is v_2. A typical path for the light to follow from S to F via the boundary is illustrated in Fig. 10.2. Let us find the condition for which the transit time t would be a minimum for such a path. The time from S to F by way of the point B on the boundary is

$$t = \frac{\overline{SB}}{v_1} + \frac{\overline{BF}}{v_1} = \frac{1}{v_1}\left[\sqrt{h_1{}^2 + x^2} + \sqrt{h_2{}^2 + (b - x)^2}\right].$$

The transit time t is a function of x. The condition for the *shortest* time is

$$\frac{dt(x)}{dx} = \frac{1}{v_1}\left[\frac{x}{\sqrt{h_1{}^2 + x^2}} - \frac{b - x}{\sqrt{h_2{}^2 + (b - x)^2}}\right]$$

$$= \frac{1}{v_1}(\sin i - \sin i') = 0$$

from which the *law of reflection* of geometrical optics follows. The angle of reflection is equal to the angle of incidence,

$$i' = i.$$

In a similar way, for light transmitted through the boundary, the transit time over a typical path from S to F via the point B on the boundary (as shown in Fig. 10.3) is

$$t(x) = \frac{\overline{SB}}{v_1} + \frac{\overline{BF}}{v_2} = \frac{1}{v_1}\sqrt{h_1{}^2 + x^2} + \frac{1}{v_2}\sqrt{h_2{}^2 + (b - x)^2}.$$

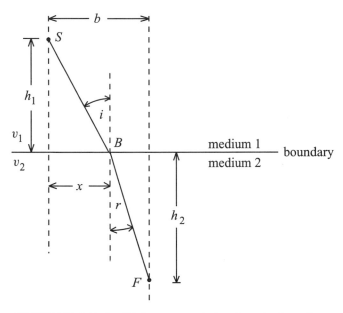

FIGURE 10.3. Path of light on transmission through a boundary.

For the path corresponding to the shortest time, we set $dt/dx = 0$, and thereby obtain Snell's law, *the law of refraction* of geometrical optics,

$$\frac{1}{v_1}\sin i = \frac{1}{v_2}\sin r.$$

In terms of the indices of refraction of the media, Snell's law is written

$$n_1 \sin i = n_2 \sin r,$$

where the index of refraction n is the ratio of the propagation velocity of light in free space (c) to the propagation velocity in the medium (v), $n = c/v$.

Thus, the laws of geometrical optics follow from a variational principle.[1]

10.2 Another Variational Calculation

In the previous section, we saw how a variational principle led to the laws of geometrical optics. Can we do something like that for other laws of physics? Mechanics, for example? To get some insight[2] on how we might formulate such a variational principle in a general way, let us consider a very simple problem: *Find the path corresponding to the shortest distance between two points in a plane.*

---

[1]A more precise statement of Fermat's principle is: *The actual path taken by the light is the one for which $\int dt$ has a* stationary *value.*

[2]See Boas, *op. cit.*, p. 386.

With the two points labeled as shown in Fig. 10.4, we see that the problem is to find the curve $y(x)$ for which the distance between the two points is minimal. The length S of the path between the two points is

$$S = \int_1^2 \sqrt{dx^2 + dy^2} = \int_{x_1}^{x_2} \sqrt{1 + \left(\frac{dy}{dx}\right)^2}\, dx = \int_{x_1}^{x_2} \sqrt{1 + y'^2}\, dx,$$

where the prime denotes differentiation with respect to x. The function $y(x)$ represents the actual (minimal) path between the two points. Now, let us consider other arbitrary paths $y(x, \alpha)$, which pass through the end points but otherwise differ slightly from $y(x)$. We label these paths with the parameter α and define them by

$$y(x, \alpha) = y(x) + \alpha \eta(x),$$

where $\eta(x)$ is an *arbitrary* function of x. We require that η vanish at x_1 and x_2. That is,

$$\eta(x_1) = \eta(x_2) = 0,$$

so that $y(x, \alpha)$ coincides with $y(x)$ at the end points.

We want to find the path $y(x, \alpha)$ for which S is a minimum (i.e., the shortest total path). Since S is different for different paths $y(x, \alpha)$, S is a function of α. That is, $S = S(\alpha)$. From the definition of $y(x, \alpha)$, we see that $y(x, 0) = y(x)$, which implies that

$$S(0) = \int_{x_1 \; (\alpha=0)}^{x_2} \sqrt{1 + y'^2}\, dx.$$

So the length of the actual path corresponds to $S(0)$, i.e., at $\alpha = 0$ where $S(\alpha)$ has its *minimum* value. Therefore, we have two related characteristics:

- $S(0) =$ the actual (shortest) path length ($\alpha = 0$),
- $S(0)$ corresponds to a *minimum* for $S(\alpha)$, hence

$$\left.\frac{dS(\alpha)}{d\alpha}\right|_{\alpha=0} = 0.$$

Carrying out the differentiation, we get

$$\frac{dS}{d\alpha} = \frac{d}{d\alpha} \int_{x_1}^{x_2} \sqrt{1 + [y'(x, \alpha)]^2}\, dx = \int_{x_1}^{x_2} \frac{\partial}{\partial \alpha} \sqrt{1 + [y'(x) + \alpha \eta'(x)]^2}\, dx$$

$$= \int_{x_1}^{x_2} \frac{\eta'(x)[y'(x) + \alpha \eta'(x)]}{\sqrt{1 + [y'(x) + \alpha \eta'(x)]^2}}\, dx = \int_{x_1}^{x_2} \frac{\eta'(x) y'(x, \alpha)}{\sqrt{1 + [y'(x, \alpha)]^2}}\, dx.$$

Now let us require that the path length $S(\alpha)$ be stationary at $\alpha = 0$,

$$\left.\frac{dS}{d\alpha}\right|_{\alpha=0} = \int_{x_1}^{x_2} \frac{y'(x)}{\sqrt{1 + (y'(x))^2}} \eta'(x)\, dx = 0.$$

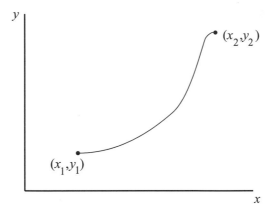

FIGURE 10.4. An arbitrary path.

Integration by parts gives two terms,

$$\frac{dS}{d\alpha}\bigg|_{\alpha=0} = \frac{y'}{\sqrt{1+y'^2}}\,\eta(x)\bigg|_{x_1}^{x_2} - \int_{x_1}^{x_2} \frac{d}{dx}\left[\frac{y'}{\sqrt{1+y'^2}}\right]\eta(x)\,dx = 0.$$

At both limits $\eta(x) = 0$, so the first term here is zero. Since $\eta(x)$ is arbitrary, the integrand in the second term must also vanish for arbitrary $\eta(x)$. Therefore,

$$\frac{d}{dx}\left[\frac{y'}{\sqrt{1+y'^2}}\right] = 0.$$

We see immediately that the quantity in the square brackets is a constant which implies that

$$y'(x) = \text{constant} \equiv c_1.$$

On integrating directly, we get

$$y = c_1 x + c_2$$

where c_1 and c_2 are constants of integration. Thus, the path we obtain for the shortest distance between two points in a plane is a *straight line*, as we already knew.

10.3   The Euler-Lagrange Equation

We can use the example in Section 10.2 as a guide to establish a general procedure for carrying out a variational calculation.[3] Common to all such calculations is that

---

[3]This development follows Boas, *op. cit.*, p. 387.

for some function $y(x)$, the integral

$$S = \int_{x_1}^{x_2} F(y, y', x)\, dx \tag{10.1}$$

is stationary. Our task is to find the path $y(x)$ for which this condition is satisfied. Following the procedure in Section 10.2, we represent an arbitrary path by

$$y(x, \alpha) = y(x) + \alpha \eta(x), \tag{10.2}$$

which coincides with the actual path $y(x)$ at $x = x_1$ and $x = x_2$ (i.e., $\eta(x_1) = \eta(x_2) = 0$). Equation (10.2) implies that $S = S(\alpha)$. We write

$$\frac{dS}{d\alpha} = \int_{x_1}^{x_2} \left[\frac{\partial F}{\partial y}\frac{\partial y}{\partial \alpha} + \frac{\partial F}{\partial y'}\frac{\partial y'}{\partial \alpha} \right] dx$$

$$= \int_{x_1}^{x_2} \left[\frac{\partial F}{\partial y}\eta(x) + \frac{\partial F}{\partial y'}\eta'(x) \right] dx.$$

We require that

$$\left. \frac{dS}{d\alpha} \right|_{\alpha=0} = \int_{x_1}^{x_2} \left[\frac{\partial F}{\partial y}\eta(x) + \frac{\partial F}{\partial y'}\eta'(x) \right] dx = 0.$$

Integrating by parts, we get

$$\left. \frac{dS}{d\alpha} \right|_{\alpha=0} = \int_{x_1}^{x_2} \frac{\partial F}{\partial y}\eta(x)\, dx + \frac{\partial F}{\partial y'}\eta(x) \bigg|_{x_1}^{x_2} - \int_{x_1}^{x_2} \frac{d}{dx}\left(\frac{\partial F}{\partial y'} \right)\eta(x)\, dx$$

$$= \int_{x_1}^{x_2} \left[\frac{\partial F}{\partial y} - \frac{d}{dx}\left(\frac{\partial F}{\partial y'} \right) \right]\eta(x)\, dx = 0,$$

where we have made use of the fact that $\eta(x)$ vanishes at both limits. Since $\eta(x)$ is arbitrary, the factor in brackets in the integrand must be zero for all values of x. That is,

$$\frac{\partial F}{\partial y} - \frac{d}{dx}\left(\frac{\partial F}{\partial y'} \right) = 0. \tag{10.3}$$

This equation is known as the *Euler-Lagrange equation*.

It is clear that to solve any problem of this type, we

- write down the integral S that is to have a stationary value,
- identify the function F in the integrand of S,
- substitute F into the Euler-Lagrange equation, and
- solve the differential equation that follows.

It is easily verified that, in finding the shortest path between two points in a plane, the procedure outlined by these four bullets leads to the same result we obtained in Section 10.2.

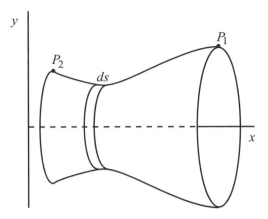

FIGURE 10.5. A surface of revolution.

10.3.1 MINIMUM AREA FOR SURFACE OF REVOLUTION

In another application,[4] we connect two points P_1 and P_2 in the xy-plane by a smooth curve and revolve the curve about the x axis to obtain a surface of revolution as shown in Fig. 10.5.

Our problem is to find the curve such that the surface has minimum area. The narrow band indicated in Fig. 10.5 has an area $2\pi y\, ds$, where ds is an element of length along the curve. Because of the way we set up the coordinate system in Fig. 10.5, it is convenient to take y to be the independent variable and x to be the dependent variable. Thus, the area of revolution is

$$\int dA = \int_1^2 2\pi y\sqrt{dx^2 + dy^2} = 2\pi \int_{y_1}^{y_2} y\sqrt{1 + x'^2}\,dy.$$

On comparison with Eq. (10.1), we identify

$$F(x, x', y) = 2\pi y\sqrt{1 + x'^2}.$$

From Eq. (10.3), we have

$$\frac{d}{dy}\left(\frac{\partial F}{\partial x'}\right) - \frac{\partial F}{\partial x} = \frac{d}{dy}\left(\frac{2\pi yx'}{\sqrt{1 + x'^2}}\right) = 0,$$

from which we see that

$$\frac{yx'}{\sqrt{1 + x'^2}} = \text{constant} \equiv c_1.$$

We solve for x' to obtain the integral

$$x = -ic_1 \int \frac{dy}{\sqrt{c_1^2 - y^2}}.$$

---

[4]G. Arfken, *Mathematical Methods for Physicists*, 2nd ed., Academic Press, New York, 1970. p. 775.

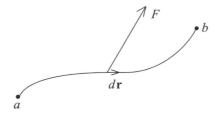

FIGURE 10.6. Work done by a force **F**.

With the change of variable, $y = c_1 \cos \theta$, we can integrate this equation directly to obtain

$$y = c_1 \cosh \left(\frac{x - c_2}{c_1} \right).$$

The constants c_1 and c_2 are fixed by the end points (x_1, y_1) and (x_2, y_2). The curve represented by $y(x)$ is called a *catenary*.

10.3.2 LAGRANGIAN FORMULATION OF MECHANICS

Now, let us use a variational principle to obtain a description of the behavior of matter moving under the influence of forces, and in particular, conservative forces as discussed in Chapter 3.

   The mechanical work W done in displacing a particle from point a to point b in a force field **F** is

$$\int_a^b dW = - \int_a^b \mathbf{F} \cdot d\mathbf{r}.$$

A typical displacement is illustrated in Fig. 10.6. If the work done is the same no matter what path we take in getting from point a to point b, then according to our work in Chapter 3, the force is conservative and expressible in terms of a function $V(\mathbf{r})$ defined at each point in space such that

$$dV = -\mathbf{F} \cdot d\mathbf{r}. \tag{10.4}$$

The value of the integral

$$\int_a^b dV = - \int_a^b \mathbf{F} \cdot d\mathbf{r}$$

depends only on the values of V at the end points of the path and not on the actual path taken. In this case, dV is an exact differential and we can write for the left-hand side of Eq. (10.4)

$$dV = \frac{\partial V}{\partial x} dx + \frac{\partial V}{\partial y} dy + \frac{\partial V}{\partial z} dz = \nabla V \cdot d\mathbf{r}.$$

The equation (10.4) may now be written as

$$(\mathbf{F} + \nabla V) \cdot d\mathbf{r} = 0,$$

which is valid for any arbitrary displacement $d\mathbf{r}$. Thus, a conservative force may be expressed as the gradient of a scalar function $V(\mathbf{r})$, which we call the potential energy, in accord with our work in Chapter 3.

$$\mathbf{F} = -\nabla V. \tag{10.5}$$

Now, consider a particle of mass m moving under the influence of conservative forces. In a time interval from t_1 to t_2, the particle moves along a path such that the integral

$$S = \int_{t_1}^{t_2} L(x, x', t)\, dt$$

is stationary along this path. The function $L(x, x', t)$, called the *Lagrangian*, is the difference between the kinetic and potential energies of the particle,

$$L(x, x', t) = \tfrac{1}{2}mx'^2 - V(x).$$

A stationary value for S requires that L satisfy the Euler-Lagrange equation,

$$\frac{d}{dt}\left(\frac{\partial L}{\partial x'}\right) - \frac{\partial L}{\partial x} = 0.$$

Substituting for L, this equation becomes

$$m\frac{d^2 x}{dt^2} + (\nabla V)_x = 0. \tag{10.6}$$

Invoking Eq. (10.5), we see that Eq. (10.6) reduces to

$$F_x = m\frac{d^2 x}{dt^2},$$

which is Newton's second law of motion. So, by a variational principle, we are led to Newton's second law. Lagrange's formulation of mechanics is equivalent to that of Newton. It is possible to base all of classical mechanics on this variational principle known as *Hamilton's principle*:

Consistent with any given constraints, a particle beginning at one point at time t_1 will move to another point at time t_2 along a path such that the integral

$$\int_{t_1}^{t_2} (T - V)\, dt$$

is stationary. In the integrand, T represents the kinetic energy of the particle and V the potential energy.

In general, the motion of a particle, and hence the Lagrangian, will depend on more than one coordinate. For n coordinates, we write

$$L = L(x_1, x_2, \ldots, x_n, x_1', x_2', \ldots, x_n', t).$$

For each coordinate x_j, there corresponds an equation

$$\frac{d}{dt}\left(\frac{\partial L}{\partial x_j'}\right) - \frac{\partial L}{\partial x_j} = 0. \tag{10.7}$$

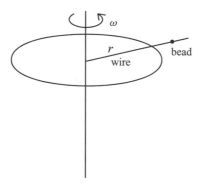

FIGURE 10.7. Bead on a rotating wire.

These are called *Lagrange's equations*.

The Lagrangian formulation proves to be a very powerful tool in quantum field theory.

Example 1.

A small bead of mass m is constrained to slide along a thin straight wire that is rotating with a constant angular velocity ω about a fixed point in a horizontal plane as illustrated in Fig. 10.7. Let r denote the distance of the bead from the axis of rotation. Find r as a function of time t.

Solution.

Choose the zero of potential energy to lie in the plane of rotation of the wire. The Lagrangian L is then just equal to the kinetic energy of the bead,

$$L = \tfrac{1}{2}m(r'^2 + r^2\omega^2).$$

With this Lagrangian, Lagrange's equation, Eq. (10.7) reduces to

$$m(r'' - \omega^2 r) = 0.$$

The general solution to this equation is

$$r(t) = Ae^{\omega t} + Be^{-\omega t}$$

where A and B are arbitrary constants. Let us assume that at time $t = 0$ the bead is at rest at a distance b from the axis of rotation. With these boundary conditions, we can evaluate the constants A and B to obtain

$$r(t) = b\cosh \omega t.$$

Example 2.

Another useful illustration of the power of variational principles is the *brachistochrone*.[5] In this problem, we consider a particle falling in a uniform gravitational

---

[5]This is a standard problem worked out in most textbooks on classical mechanics. See, for example Marion, *op. cit.*, p. 184.

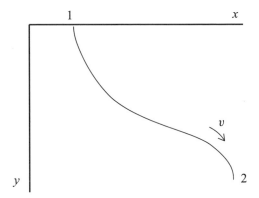

FIGURE 10.8. The brachistochrone.

field. What is the path along which the particle starting from rest will fall from point 1 to point 2 in the *shortest* time? (See Fig. 10.8.)

Solution.

The problem is to find the path for which $\int dt$ is a minimum. First, choose the potential energy to be zero at point 1. From the principle of energy conservation, we then find that when the particle has vertical position y its speed is $v = \sqrt{2gy}$, which leads to

$$\int dt = \int_1^2 \frac{ds}{\sqrt{2gy}}$$

where ds is an infinitesimal length along the path. Since x is missing in the integrand, we write $ds = \sqrt{1 + x'^2}\, dy$. This gives

$$\int dt = \frac{1}{\sqrt{2g}} \int_{y_1}^{y_2} \sqrt{\frac{1 + x'^2}{y}}\, dy.$$

Clearly,

$$F(x, x', y) = \sqrt{\frac{1 + x'^2}{y}}.$$

Substituting this function into the Euler-Lagrange equation, Eq. (10.3), yields

$$\frac{x'}{\sqrt{y(1 + x'^2)}} = \text{constant} \equiv c_1.$$

We solve this equation for x' and integrate to obtain

$$x = 2c_1^2 \int \frac{y\, dy}{\sqrt{1 - (1 - 2c_1^2 y)^2}}.$$

With the change of variable, $\cos \theta = 1 - 2c_1^2 y$ or

$$y = \frac{1}{2c_1^2}(1 - \cos \theta), \qquad (10.8)$$

we get

$$x = \frac{1}{2c_1^2}(\theta - \sin \theta) + c_2. \qquad (10.9)$$

Equations (10.8) and (10.9) provide a parametric representation of a curve known as a *cycloid*.[6] If a circular disk is rolled without slipping over a flat surface, a point fixed on the rim of the disk traces out a cycloid. If the cycloid passes through the origin, where $x = y = 0$, we find that

$$x = -\sqrt{\frac{y}{c_1^2} - y^2} + \frac{1}{2c_1^2} \cos^{-1}(1 - 2c_1^2 y).$$

10.3.3 CONSTRAINTS

In addition to the requirement that an integral have a stationary value in problems involving the variational calculus, there may exist other relationships that constrain the variables and must be satisfied. This is implicit in the statement of Hamilton's principle on page 215.

To illustrate this point, we consider a uniform cable of length l and mass m. The ends of the cable are attached to fixed points such that the cable is suspended in equilibrium in the earth's gravitational field as illustrated in Fig. 10.9. The problem is to find the shape of the hanging cable.

With the cable in equilibrium, the gravitational potential energy is a minimum. Therefore, we should like to find the relation between the vertical position y and

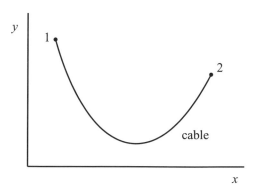

FIGURE 10.9. A uniform hanging cable.

---

[6]Compare these results with Eqs. (5.21) and (5.22).

the horizontal position x for all points along the cable such that the gravitational potential energy given by

$$V = \frac{mg}{l} \int_1^2 y \, ds$$

is a minimum. That is, we require that the integral

$$\int_1^2 y \, ds \tag{10.10}$$

be stationary.

There is a problem here that we haven't encountered in the previous examples. The various paths $y(\alpha, x)$ are not completely arbitrary, but must be subject to the condition that the total length of the path remain constant. The length of the path must be equal to the fixed length l of the cable,

$$\int_1^2 ds = l. \tag{10.11}$$

Both requirements, a stationary value for the integral in Eq. (10.10) and the constraint in Eq. (10.11), can be incorporated in a single equation. From Eq. (10.11), it follows that

$$\lambda \left(\int_1^2 ds - l \right) = 0, \tag{10.12}$$

where λ is an arbitrary constant. Since l is fixed, we see from Eq. (10.12) that the quantity

$$\lambda \int_1^2 ds \tag{10.13}$$

must also be stationary for the actual path.

Combining Eqs. (10.10) and (10.13), we require that the integral

$$\int_1^2 (y + \lambda) \, ds = \int_{y_1}^{y_2} (y + \lambda) \sqrt{1 + x'^2} \, dy$$

have a stationary value. The constraint of Eq. (10.11) is now represented by the parameter λ. The corresponding Euler-Lagrange equation is

$$\frac{d}{dy} \left[\frac{\partial}{\partial x'} \left((y + \lambda) \sqrt{1 + x'^2} \right) \right] - \frac{\partial}{\partial x} \left((y + \lambda) \sqrt{1 + x'^2} \right) = \frac{d}{dy} \left[\frac{(y + \lambda) x'}{\sqrt{1 + x'^2}} \right] = 0.$$

Thus,

$$\frac{(y + \lambda) x'}{\sqrt{1 + x'^2}} = c_1. \tag{10.14}$$

On rewriting Eq. (10.14), we obtain an expression for $x(y)$,

$$x(y) = c_1 \int \frac{dy}{\sqrt{(y + \lambda)^2 - c_1^2}} + c_2,$$

where c_1 and c_2 are integration constants. Performing the integration, we have

$$x(y) = c_1 \log\left(\frac{y+\lambda}{c_1} + \sqrt{\left(\frac{y+\lambda}{c_1}\right)^2 - 1}\right) + c_2.$$

Solving for $y(x)$,

$$y(x) = -\lambda + c_1 \cosh\left(\frac{x - c_2}{c_1}\right), \tag{10.15}$$

with three parameters λ, c_1, and c_2 to be determined from the constraint in Eq. (10.11) and the two fixed end points.

For the derivative of $y(x)$, we get

$$y'(x) = \sinh\left(\frac{x - c_2}{c_1}\right).$$

Invoking the constraint in Eq. (10.11) yields

$$\int_1^2 ds = \int_{x_1}^{x_2} \sqrt{1 + y'^2}\,dx$$

$$= \int_{x_1}^{x_2} \sqrt{1 + \sinh^2\left(\frac{x - c_2}{c_1}\right)}\,dx = \int_{x_1}^{x_2} \cosh\left(\frac{x - c_2}{c_1}\right)dx$$

$$= c_1\left(\sinh\left(\frac{x_2 - c_2}{c_1}\right) - \sinh\left(\frac{x_1 - c_2}{c_1}\right)\right) = l,$$

from which we could obtain c_2 in terms of c_1 and the given parameters x_1, x_2, and l. By requiring the curve to pass through the endpoints of the cable at (x_1, y_1) and (x_2, y_2) the other two constants c_1 and λ are uniquely determined. From Eq. (10.15), we see that the shape of the hanging cable is a catenary.

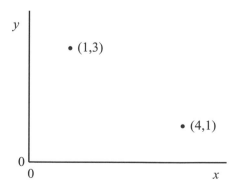

FIGURE 10.10. Coordinate system for Exercise 10.1.

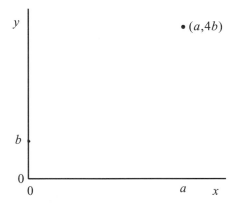

FIGURE 10.11. Coordinate system for Exercise 10.5.

10.4   Exercises

1. Light propagates in a medium where the index of refraction n varies with coordinate y according to

$$n = \frac{n_0}{y}.$$

Use Fermat's principle to find the path of the light through the points with coordinates (x, y) shown in Fig. 10.10. Describe clearly and completely in words the path of the light.

2. Write down the Euler-Lagrange equation for which the integral

$$\int_1^2 \sqrt{y}\, ds$$

is stationary. Here 1 and 2 correspond to two points in the xy-plane and $ds = \sqrt{dx^2 + dy^2}$. Solve the Euler-Lagrange equation to obtain y as an explicit function of x.

3. The integral

$$\int_{x_1}^{x_2} (y^2 - y'^2)\, dx$$

has a stationary value along the path $y = y(x)$ which passes through the points $(\frac{\pi}{4}, 0)$ and $(0, 1)$ in the xy-plane. Use the calculus of variations to find $y(x)$.

4. Find the curve $y = y(x)$ in the xy-plane passing through the points $(-6, 12)$ and $(4, 27)$ such that the integral

$$\int_{x_1}^{x_2} (x + y + y'^2)\, dx$$

has a stationary value.

5. Require that the integral

$$\int_{x_1}^{x_2} \frac{y'^2}{x} \, dx$$

be stationary. From the Euler-Lagrange equation, find the corresponding path passing through the points $(0, b)$ and $(a, 4b)$ as shown in Fig. 10.11. Make a sketch of the path in the xy-plane.

6. Find the path in the xy-plane along which the integral

$$\int_{x_1}^{x_2} x(1 - x)y'^3 \, dx$$

is stationary. This path passes through the origin and the point $(1, \frac{2}{3})$. Write x explicitly as a function of y with the appropriate values for the constants of integration.

7. Points in the plane of propagation of light in a certain medium are represented by the rectangular coordinates x and y. The index of refraction of this medium depends on the position according to

$$n = \frac{n_0}{b} y,$$

where n_0 and b are real, positive constants. Use Fermat's principle to obtain the path the light takes in traveling from the point $(0, b)$ to the point $(b, b \cosh 1)$. Express your result in terms of y as an explicit function of x.

8. A particle of mass m is constrained to move along the x-axis. Its potential energy is

$$PE = \tfrac{1}{2}m\omega^2 x^2$$

where ω is a real constant. Invoke Hamilton's principle and obtain Lagrange's equation for this particle. At time $t = 0$ the particle is moving to the right with speed v at a distance a to the right of the origin. Find the position x as an explicit function of t. Describe in words the behavior of this particle.

9. A simple pendulum consists of a point mass m attached to one end of a massless rod of length l. The other end of the rod is fixed at a point P. The pendulum is free to swing in a vertical plane about the fixed point as shown in Fig. 10.12. Write down the Lagrangian for this system. Explain your reasoning clearly and completely in words. From Lagrange's equation, obtain the equation of motion (a differential equation in θ and t) for the mass m.

10. A particle of mass m moves under the influence of a conservative force that depends only on the distance r of the particle from the force center. The potential energy of this particle is $U(\mathbf{r}) = U(r)$. Starting with Eq. (2.13), show that the angular momentum $\mathbf{L} = \mathbf{r} \times \mathbf{p}$ of this particle is conserved. Use this result to show that the motion of the particle is confined to a fixed plane. Hence, only two independent coordinates are required to describe the motion. Choose these to be r and the angle θ that gives the angular position of the particle relative to some arbitrary axis lying in the plane of motion and

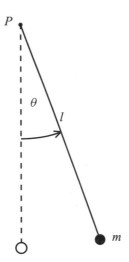

FIGURE 10.12. A simple pendulum for Exercise 10.9.

passing through the force center. Write the Lagrangian for this system and obtain Lagrange's equations. Show that one of these equations is a reflection of the fact that angular momentum is conserved.

11. One end of a light, inextensible string of length l is fastened to a particle of mass m. The other end of the string is attached to a second particle of mass M. The mass M is constrained to slide on a smooth, horizontal surface. The string passes through a hole in the surface such that the mass m hangs directly below the hole without swinging as illustrated in Fig. 10.13. Construct the Lagrangian for this two-particle system. Write down Lagrange's equation for each of the two variables r (the distance from the hole to M) and θ. Obtain the equations of motion for arbitrary r and θ. For the special case in which the mass M describes a circular orbit with constant angular frequency ω, show that the radius of this orbit is $mg/M\omega^2$.

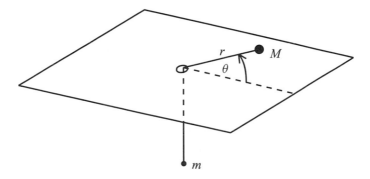

FIGURE 10.13. Two masses connected by a string for Exercise 10.11.

12. A smooth wire is bent in the shape of a circle of radius R. A small bead of mass m is free to slide on the wire. This circular loop lies in a vertical plane and rotates with a constant angular velocity ω about a vertical axis passing through the center of the loop. A Lagrangian for this particle is given by

$$L = \tfrac{1}{2}mR^2(\theta'^2 + \omega^2 \sin^2 \theta) - mgR(1 + \cos \theta),$$

where θ represents the angular position of the particle on the wire. Draw a diagram illustrating this arrangement with the relevant parameters clearly identified. Indicate on the drawing how θ and the zero of potential energy must be chosen to be consistent with the above Lagrangian. From this Lagrangian, obtain the differential equation that describes the behavior of θ in time. Discuss the motion of the particle for the special case when $\theta(t) = \pi - \varepsilon(t)$ with $|\varepsilon(t)| \ll 1$. Consider separately the two situations when the wire is rotating rapidly ($\omega > \sqrt{g/R}$) and when it is rotating slowly ($\omega < \sqrt{g/R}$).

13. The Lagrangian L for a particle of mass m is given by

$$L = \tfrac{1}{2}mv^2 - \tfrac{1}{2}kx^2 + \gamma x^3,$$

where x and v represent the particle's position and velocity, respectively. The parameters k and γ are real, positive constants. From this Lagrangian, obtain the equation of motion for the particle.

14. Two particles of mass m and M with $m < M$, are connected by a light, inextensible string. The string is passed over a pulley of negligible mass so that the particles are suspended in the earth's gravitational field as illustrated in Fig. 10.14. The radius of the pulley is R and the length of the string is $40R$. (The drawing is not to scale.) The pulley is free to rotate on its frictionless bearing. Static friction between the pulley and the string causes the string to move without slipping as the pulley rotates. The vertical position of the mass M relative to the top of the pulley is denoted by x. Construct the Lagrangian $L = L(x, x', t)$ for this system and from it obtain the equation of motion. Assume that at time $t = 0$, the position x and velocity v of the mass M are

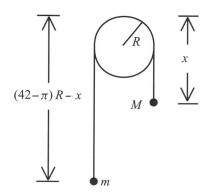

$(42 - \pi)R - x$

FIGURE 10.14. An Atwood's machine for Exercise 10.14.

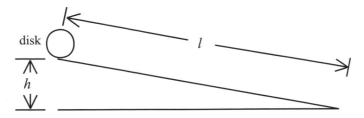

FIGURE 10.15. Disk on incline for Exercise 10.15.

given by

$$x(0) = 4R,$$

$$v(0) = 2\sqrt{\frac{(M-m)gR}{M+m}}.$$

From your equation of motion, obtain an expression for x as an explicit function of t such that these initial conditions are satisfied. Use your result to calculate the time at which the smaller mass m is at a distance $(8 - \pi)R$ below the top of the pulley.

15. A uniform disk of mass M and radius R starts from rest at the top of an incline and rolls without slipping down the incline. The height of the incline is denoted by h and the length by l as indicated in Fig. 10.15. Write down the Lagrangian for this disk. From the Lagrangian, obtain the equation of motion and solve it subject to the initial conditions. From your result, show that the time it takes for the disk to roll half way down the incline is $\sqrt{3l^2/2gh}$.

16. The mass m of a rapidly moving particle depends on its speed v according to

$$m = \frac{m_0}{\sqrt{1 - \dfrac{v^2}{c^2}}},$$

where m_0 is the mass of the particle at rest and c is the speed of light in free space. If the linear momentum $\mathbf{p}$ of the particle is written in the usual form $\mathbf{p} = m\mathbf{v}$ with m now the relativistic mass given above, then Newton's second law written in the form

$$\mathbf{F} = \frac{d\mathbf{p}}{dt},$$

is still valid. For a particle of rest mass m_0 acted on by a conservative, velocity-independent force represented by the potential energy $V(\mathbf{r})$, a relativistic Lagrangian is given by

$$L = -m_0 c^2 \sqrt{1 - \frac{v^2}{c^2}} - V(\mathbf{r}).$$

Show that this Lagrangian leads to Newton's second law in the form given above.

17. Two of Maxwell's equations[7] of electrodynamics are

$$\nabla \cdot \mathbf{B} = 0 \qquad \text{and} \qquad \nabla \times \mathbf{E} + \frac{\partial \mathbf{B}}{\partial t} = 0,$$

where $\mathbf{E}$ and $\mathbf{B}$ are the electric and magnetic fields, respectively. Using vector identities[8] show that both equations are satisfied if we define the fields in terms of a scalar $U(\mathbf{r}, t)$ and a vector $\mathbf{A}(\mathbf{r}, t)$ according to

$$\mathbf{B} = \nabla \times \mathbf{A} \qquad \text{and} \qquad \mathbf{E} = -\nabla U - \frac{\partial \mathbf{A}}{\partial t}.$$

A nonrelativistic particle of mass m and electric charge e is moving with velocity $\mathbf{v}$ in an electromagnetic field. The Lagrangian[9] is

$$L = \tfrac{1}{2}m\mathbf{v}^2 + e(\mathbf{v} \cdot \mathbf{A} - U),$$

where U and $\mathbf{A}$ are functions of space and time. Obtain the corresponding equations of motion from Lagrange's equations. (Do *not* assume the particle velocity to be constant.) Identify your result with Newton's second law and show explicitly that the electromagnetic force on the particle implied by the Lagrangian L is

$$\mathbf{F}_{\text{em}} = e(\mathbf{E} + \mathbf{v} \times \mathbf{B}).$$

18. Let $S = S(\mathbf{r}, t)$ be any scalar function of position $\mathbf{r}$ and time t and make the following replacements in the Lagrangian L for the particle in an electromagnetic field in Exercise 10.17,

$$\mathbf{A} \longrightarrow \mathbf{A} + \nabla S \qquad \text{and} \qquad U \longrightarrow U - \frac{\partial S}{\partial t}.$$

Show that Lagrange's equations are not affected by this transformation. Show directly that the electric and magnetic fields also remain unchanged by this transformation.

19. Suppose the Lagrangian for a certain particle with velocity $\mathbf{v}$ can be written as

$$L = \alpha(\mathbf{r}) \cdot \mathbf{v} - \beta(\mathbf{r}),$$

where α and β are explicit functions of position $\mathbf{r}$ but not time t. Show that Lagrange's equations lead to the relation

$$\mathbf{v} \times (\nabla \times \alpha) = \nabla \beta.$$

20. There are physical similarities in the brachistochrone problem of Example 2 and the particle moving in the uniform electric and magnetic fields described in Section 5.4.2. Identify these similarities and explain why the paths of the particles are similar.

---

[7] See Appendix B.
[8] See Appendix A.
[9] H. Goldstein, *Classical Mechanics*, Addison-Wesley, Reading MA, 1950, p. 23.

Appendix A
Vector Relations

A.1 Vector Identities

$$\mathbf{a} \cdot (\mathbf{b} \times \mathbf{c}) = (\mathbf{a} \times \mathbf{b}) \cdot \mathbf{c}$$

$$\mathbf{a} \times (\mathbf{b} \times \mathbf{c}) = (\mathbf{a} \cdot \mathbf{c})\mathbf{b} \quad (\mathbf{a} \cdot \mathbf{b})\mathbf{c}$$

$$\nabla(\mathbf{F} \cdot \mathbf{G}) = (\mathbf{G} \cdot \nabla)\mathbf{F} + (\mathbf{F} \cdot \nabla)\mathbf{G} + \mathbf{G} \times (\nabla \times \mathbf{F}) + \mathbf{F} \times (\nabla \times \mathbf{G})$$

$$\nabla \cdot (\psi\mathbf{F}) = \psi\nabla \cdot \mathbf{F} + \mathbf{F} \cdot \nabla\psi$$

$$\nabla \times (\psi\mathbf{F}) = \psi\nabla \times \mathbf{F} + (\nabla\psi) \times \mathbf{F}$$

$$\nabla \cdot (\mathbf{F} \times \mathbf{G}) = \mathbf{G} \cdot (\nabla \times \mathbf{F}) - \mathbf{F} \cdot (\nabla \times \mathbf{G})$$

$$\nabla \times (\mathbf{F} \times \mathbf{G}) = (\mathbf{G} \cdot \nabla)\mathbf{F} + \mathbf{F}(\nabla \cdot \mathbf{G}) - (\mathbf{F} \cdot \nabla)\mathbf{G} - \mathbf{G}(\nabla \cdot \mathbf{F})$$

$$\nabla \cdot (\nabla \times \mathbf{F}) = 0$$

$$\nabla \times (\nabla \times \mathbf{F}) = -\nabla^2\mathbf{F} + \nabla(\nabla \cdot \mathbf{F})$$

$$\nabla \cdot (\nabla\psi) = \nabla^2\psi$$

$$\nabla \times (\nabla\psi) = 0$$

A.2 Integral Theorems

$$\text{Divergence theorem}: \quad \oint_S \mathbf{F} \cdot d\mathbf{A} = \int_V \nabla \cdot \mathbf{F}\, d^3r$$

$$\text{Stokes's theorem}: \quad \oint_C \mathbf{F} \cdot d\mathbf{l} = \int_S (\nabla \times \mathbf{F}) \cdot d\mathbf{\Lambda}$$

A.3   The Functions of Vector Calculus

A.3.1   RECTANGULAR COORDINATES

$$\nabla \psi = \mathbf{i}\frac{\partial \psi}{\partial x} + \mathbf{j}\frac{\partial \psi}{\partial y} + \mathbf{k}\frac{\partial \psi}{\partial z}$$

$$\nabla \cdot \mathbf{F} = \frac{\partial F_x}{\partial x} + \frac{\partial F_y}{\partial y} + \frac{\partial F_z}{\partial z}$$

$$\nabla \times \mathbf{F} = \mathbf{i}\left(\frac{\partial F_z}{\partial y} - \frac{\partial F_y}{\partial z}\right) + \mathbf{j}\left(\frac{\partial F_x}{\partial z} - \frac{\partial F_z}{\partial x}\right) + \mathbf{k}\left(\frac{\partial F_y}{\partial x} - \frac{\partial F_x}{\partial y}\right)$$

$$\nabla^2 \psi = \frac{\partial^2 \psi}{\partial x^2} + \frac{\partial^2 \psi}{\partial y^2} + \frac{\partial^2 \psi}{\partial z^2}$$

A.3.2   SPHERICAL POLAR COORDINATES

$$\nabla \psi = \hat{\mathbf{r}}\frac{\partial \psi}{\partial r} + \hat{\boldsymbol{\theta}}\frac{1}{r}\frac{\partial \psi}{\partial \theta} + \hat{\boldsymbol{\phi}}\frac{1}{r \sin \theta}\frac{\partial \psi}{\partial \phi}$$

$$\nabla \cdot \mathbf{F} = \frac{1}{r^2}\frac{\partial}{\partial r}(r^2 F_r) + \frac{1}{r \sin \theta}\frac{\partial}{\partial \theta}(\sin \theta F_\theta) + \frac{1}{r \sin \theta}\frac{\partial F_\phi}{\partial \phi}$$

$$\nabla \times \mathbf{F} = \hat{\mathbf{r}}\frac{1}{r \sin \theta}\left(\frac{\partial}{\partial \theta}(\sin \theta F_\phi) - \frac{\partial F_\theta}{\partial \phi}\right)$$
$$+ \hat{\boldsymbol{\theta}}\frac{1}{r}\left(\frac{1}{\sin \theta}\frac{\partial F_r}{\partial \phi} - \frac{\partial}{\partial r}(r F_\phi)\right) + \hat{\boldsymbol{\phi}}\frac{1}{r}\left(\frac{\partial}{\partial r}(r F_\theta) - \frac{\partial F_r}{\partial \theta}\right)$$

$$\nabla^2 \psi = \frac{1}{r^2}\frac{\partial}{\partial r}\left(r^2\frac{\partial \psi}{\partial r}\right) + \frac{1}{r^2 \sin \theta}\frac{\partial}{\partial \theta}\left(\sin \theta\frac{\partial \psi}{\partial \theta}\right) + \frac{1}{r^2 \sin^2 \theta}\frac{\partial^2 \psi}{\partial \phi^2}$$

A.3.3   CYLINDRICAL COORDINATES

$$\nabla \psi = \hat{\mathbf{r}}\frac{\partial \psi}{\partial r} + \hat{\boldsymbol{\phi}}\frac{1}{r}\frac{\partial \psi}{\partial \phi} + \mathbf{k}\frac{\partial \psi}{\partial z}$$

$$\nabla \cdot \mathbf{F} = \frac{1}{r}\frac{\partial}{\partial r}(r F_r) + \frac{1}{r}\frac{\partial F_\phi}{\partial \phi} + \frac{\partial F_z}{\partial z}$$

$$\nabla \times \mathbf{F} = \hat{\mathbf{r}}\left(\frac{1}{r}\frac{\partial F_z}{\partial \phi} - \frac{\partial F_\phi}{\partial z}\right)$$
$$+ \hat{\boldsymbol{\phi}}\left(\frac{\partial F_r}{\partial z} - \frac{\partial F_z}{\partial r}\right) + \mathbf{k}\frac{1}{r}\left(\frac{\partial}{\partial r}(r F_\phi) - \frac{\partial F_r}{\partial \phi}\right)$$

$$\nabla^2 \psi = \frac{1}{r}\frac{\partial}{\partial r}\left(r\frac{\partial \psi}{\partial r}\right) + \frac{1}{r^2}\frac{\partial^2 \psi}{\partial \phi^2} + \frac{\partial^2 \psi}{\partial z^2}$$

Appendix B
Fundamental Equations of Physics

B.1   Poisson's Equation

$$\nabla^2 U(\mathbf{r}) = -\rho(\mathbf{r})$$

B.2   Laplace's Equation

$$\nabla^2 U(\mathbf{r}) = 0$$

B.3   Maxwell's Equations

$$\nabla \cdot \mathbf{D} = \rho(\mathbf{r}, t) \qquad \text{(Gauss's law)}$$
$$\nabla \cdot \mathbf{B} = 0$$
$$\nabla \times \mathbf{E} + \frac{\partial \mathbf{B}}{\partial t} = 0 \qquad \text{(Faraday's law)}$$
$$\nabla \times \mathbf{H} - \frac{\partial \mathbf{D}}{\partial t} = \mathbf{j}(\mathbf{r}, t) \qquad \text{(Ampere's law)}$$

with the auxiliary relations $\mathbf{D} = \epsilon \mathbf{E}$ and $\mathbf{B} = \mu \mathbf{H}$

B.4   Time-Dependent Schrödinger Equation

$$\left[\frac{-\hbar^2}{2m}\frac{\partial^2}{\partial x^2} + V(x)\right]\psi(x,t) = i\hbar\frac{\partial}{\partial t}\psi(x,t) \qquad \text{(one dimension)}$$

$$\left[\frac{-\hbar^2}{2m}\nabla^2 + V(\mathbf{r})\right]\psi(\mathbf{r},t) = i\hbar\frac{\partial}{\partial t}\psi(\mathbf{r},t) \qquad \text{(three dimensions)}$$

Appendix C
Some Useful Integrals and Sums

In the following expressions, m and n are nonnegative integers and a and b are constants.

C.1   Integrals

$$\int_0^\infty t^{2n} e^{-a^2 t^2} \, dt = \frac{(2n)! \sqrt{\pi}}{(2a)^{2n+1} n!}$$

$$\int_0^\infty t^{2n+1} e^{-a^2 t^2} \, dt = \frac{n!}{2a^{2n+2}}$$

$$\int_0^\infty t^n e^{-at} \, dt = \frac{n!}{a^{n+1}}$$

$$\int_0^\infty \frac{t^m}{(1+t)^{m+n+2}} \, dt = \frac{m! n!}{(m+n+1)!}$$

$$\int_0^1 t^m (1-t)^n \, dt = \frac{m! \, n!}{(m+n+1)!}$$

$$\int_{-1}^1 \frac{dy}{y-x} = \log\left(\frac{x-1}{x+1}\right) \quad \text{for} \quad |x| > 1.$$

$$\int_{-1}^1 \frac{y \, dy}{y-x} = 2 + x \log\left(\frac{x-1}{x+1}\right) \quad \text{for} \quad |x| > 1.$$

$$\int_{-1}^1 \frac{y^n \, dy}{y-x} = \frac{n(x+1) - 1 + (-1)^n [n(x-1) + 1]}{n(n-1)} + x^2 \int_{-1}^1 \frac{y^{n-2} \, dy}{y-x}$$
$$\text{for} \quad |x| > 1 \quad \text{and} \quad n \geq 2.$$

$$\int \frac{dx}{\sqrt{x^2 \pm a^2}} = \log\left[\frac{x + \sqrt{x^2 \pm a^2}}{a}\right]$$

$$\int \frac{x \, dx}{\sqrt{a^2 + x^2}} = \sqrt{a^2 + x^2}$$

$$\int \frac{x^3 \, dx}{\sqrt{a^2 + x^2}} = \frac{1}{3}\sqrt{a^2 + x^2} \, (x^2 - 2a^2)$$

$$\int_0^{2\pi} \sin mx \sin nx \, dx = \begin{cases} \pi \delta_{mn} & \text{for } m \neq 0 \\ 0 & \text{for } m = 0 \text{ or } n = 0, \end{cases}$$

$$\int_0^{2\pi} \cos mx \cos nx \, dx = \begin{cases} \pi \delta_{mn} & \text{for } m \neq 0 \\ 2\pi & \text{for } m = n = 0 \end{cases}$$

$$\int_0^{2\pi} \sin mx \cos nx \, dx = 0$$

$$\int_0^{2\pi} \sin \frac{mx}{2} \sin \frac{nx}{2} \, dx = \begin{cases} \pi \delta_{mn} & \text{for } m \neq 0 \\ 0 & \text{for } m = 0 \text{ or } n = 0, \end{cases}$$

$$\int_0^{2\pi} \cos \frac{mx}{2} \cos \frac{nx}{2} \, dx = \begin{cases} \pi \delta_{mn} & \text{for } m \neq 0 \\ 2\pi & \text{for } m = n = 0 \end{cases}$$

$$\int \sin^2 dx = \tfrac{1}{2}\left(x - \tfrac{1}{2}\sin 2x\right)$$

$$\int \cos^2 dx = \tfrac{1}{2}(x + \tfrac{1}{2}\sin 2x)$$

$$\int \sec x \, dx = \log\left(\sec x + \tan x\right)$$

$$\int \cos^n x \sin x \, dx = \frac{-1}{n+1} \cos^{n+1} x$$

$$\int \sin^n x \cos x \, dx = \frac{1}{n+1} \sin^{n+1} x$$

$$\int \frac{dx}{1 + \cos x} = \tan \frac{x}{2}$$

$$\int \sin ax \sin bx \, dx = \frac{1}{2}\left[\frac{\sin[(a-b)x]}{a-b} - \frac{\sin[(a+b)x]}{a+b} \right]$$

$$\int \cos ax \cos bx \, dx = \frac{1}{2}\left[\frac{\sin[(a-b)x]}{a-b} + \frac{\sin[(a+b)x]}{a+b} \right]$$

$$\int \sin ax \cos bx \, dx = -\frac{1}{2}\left[\frac{\cos[(a-b)x]}{a-b} + \frac{\cos[(a+b)x]}{a+b} \right]$$

C.2 Sums

$$\sum_{n=1}^{\infty} \frac{1}{n^2} = \frac{\pi^2}{6} \qquad\qquad \sum_{n=1}^{\infty} \frac{1}{n^4} = \frac{\pi^4}{90}$$

$$\sum_{n=1}^{\infty} \frac{1}{n^6} = \frac{\pi^6}{945} \qquad\qquad \sum_{n=1}^{\infty} \frac{1}{n^8} = \frac{\pi^8}{9450}$$

$$\sum_{n=1}^{\infty} \frac{(-1)^n}{n^2} = -\frac{\pi^2}{12} \qquad\qquad \sum_{n=1}^{\infty} \frac{(-1)^n}{n^4} = -\frac{7\pi^4}{720}$$

$$\sum_{n=0}^{\infty} \frac{1}{(2n+1)^2} = \frac{\pi^2}{8} \qquad\qquad \sum_{n=0}^{\infty} \frac{1}{(2n+1)^4} = \frac{\pi^4}{96}$$

$$\sum_{n=0}^{\infty} \frac{(-1)^n}{2n+1} = \frac{\pi}{4} \qquad\qquad \sum_{n=0}^{\infty} \frac{(-1)^n}{(2n+1)^3} = \frac{\pi^3}{32}$$

Appendix D
Algebraic Equations

D.1 Quadratic Equation

It is well known that any quadratic algebraic equation can be expressed in the form

$$x^2 + bx + c = 0, \tag{D.1}$$

which has solutions

$$x = \frac{-b \pm \sqrt{b^2 - 4c}}{2}. \tag{D.2}$$

D.2 Cubic Equation

The general cubic algebraic equation can be written,

$$x^3 + ax^2 + bx + c = 0. \tag{D.3}$$

For our purpose, we assume the coefficients a, b, and c are all real. By a sequence of redefinitions of the unknown quantity, the solution of this equation can be reduced to the solution of a quadratic equation, as we shall now show.[1]

First, define a new unknown according to $x = y + s$ and choose s so that the resulting cubic equation in y has no y^2 term. With this substitution in Eq. (D.3)

---

[1] See, for example, Raymond W. Brink, *College Algebra*, 2nd ed., Appleton-Century-Crofts, Inc., New York, 1951, p. 333.

we have,

$$y^3 + (3s + a)y^2 + (3s^2 + 2as + b)y + s^3 + as^2 + bs + c = 0.$$

Choosing $s = -a/3$, reduces this equation to

$$y^3 + ty + u = 0, \tag{D.4}$$

where we have defined $t = b - a^2/3$ and $u = (2a^3 - 9ab + 27c)/27$. The next step is to express y in terms of a new unknown z so that Eq. (D.4) is transformed into a quadratic equation. For this purpose, we write $y = z + v/z$. With this substitution for y, Eq. (D.4) becomes,

$$z^6 + (3v + t)z^4 + uz^3 + v(3v + t)z^2 + v^3 = 0.$$

Choosing $v = -t/3$ reduces this equation to

$$z^6 + uz^3 - \frac{t^3}{27} = 0, \tag{D.5}$$

which is a quadratic equation for z^3. From Eq. (D.2), we see that the solution is

$$z^3 = \frac{-u + \sqrt{u^2 + \frac{4t^3}{27}}}{2}, \tag{D.6}$$

where we have taken the upper sign in Eq. (D.2). Equation (D.6) is of the form $z^3 = a$ discussed in Section 4.2 and, in general, has three solutions,

$$z = \left| \frac{-u + \sqrt{u^2 + \frac{4t^3}{27}}}{2} \right|^{\frac{1}{3}} e^{i(\phi + 2m\pi)/3} \qquad \text{with} \quad m = 0, 1, 2. \tag{D.7}$$

The phase ϕ is defined by,

$$\frac{-u + \sqrt{u^2 + \frac{4t^3}{27}}}{2} = \left| \frac{-u + \sqrt{u^2 + \frac{4t^3}{27}}}{2} \right| e^{i\phi}. \tag{D.8}$$

Working backward from Eq. (D.7), we use our results for z to construct y (of Eq. (D.4)), and from y we obtain x, the solutions to the original cubic equation, Eq. (D.3).

References

1. G. Arfken, *Mathematical Methods for Physicists*, 2nd ed., Academic Press, London, 1970.
2. M. L. Boas, *Mathematical Methods in the Physical Sciences*, 2nd ed., John Wiley and Sons, New York, 1983.
3. P. A. M. Dirac, *The Principles of Quantum Mechanics*, Oxford University Press, Oxford, 1928.
4. A. L. Fetter and J. D. Walecka, *Theoretical Mechanics of Particles and Continua*, McGraw-Hill Book Company, New York, 1980.
5. H. Goldstein, *Classical Mechanics*, Addison-Wesley, Reading, MA, 1950.
6. D. J. Griffiths, *Introduction to Electrodynamics*, Prentice-Hall, Englewood Cliffs, NJ, 1981.
7. M. Jammer, *The Conceptual Development of Quantum Mechanics*, McGraw-Hill Book Company, New York, 1966.
8. W. Kaplan, *Advanced Calculus*, Addison-Wesley, Reading, MA, 1953.
9. H. Margenau and G. M. Murphy, *The Mathematics of Physics and Chemistry*, Van Nostrand, Princeton, 1956.
10. J. B. Marion, *Classical Dynamics of Particles and Systems*, 2nd ed., Academic Press, New York, 1970.
11. P. T. Matthews, *Introduction to Quantum Mechanics*, 3rd ed., McGraw-Hill Book Company (UK), Ltd., London, 1974.
12. E. Merzbacher, *Quantum Mechanics*, John Wiley & Sons, New York, 1961.
13. J. B. Seaborn, *Hypergeometric Functions and Their Applications*, Springer-Verlag, New York, 1991.
14. E. C. Titchmarsh, *The Theory of Functions*, Oxford University Press, Oxford, 1939.

Bibliography

1. G. Arfken and H. J. Weber, *Mathematical Methods for Physicists*, 5th ed., Harcourt/Academic Press, San Diego, 2001.
2. S. Axler, *Linear Algebra Done Right*, Springer-Verlag, New York, 1996.
3. V. D. Barger and M. G. Olsson, *Classical Mechanics: A Modern Perspective*, 2nd ed., McGraw-Hill, Inc., 1995.
4. R. A. Becker, *Introduction to Theoretical Mechanics*, McGraw-Hill Book Company, New York, 1954.
5. R. W. Brink, *College Algebra*, 2nd ed., Appleton-Century-Crofts, Inc., New York, 1951.
6. R. V. Churchill, *Fourier Series and Boundary Value Problems*, McGraw-Hill Book Company, New York, 1941.
7. H. Jeffreys and B. Jeffreys, *Methods of Mathematical Physics*, Cambridge University Press, Cambridge, U.K. 1956.
8. P. B. Kahn, *Mathematical Methods for Scientists and Engineers: Linear and Nonlinear Systems*, John Wiley and Sons, New York, 1990.
9. E. A. Kraut, *Fundamentals of Mathematical Physics*, McGraw-Hill Book Company, New York, 1972.
10. N. N. Lebedev, *Special Functions and Their Applications*, Dover Publications, Inc., New York, 1972.
11. J. D. Logan, *Applied Partial Differential Equations*, Springer-Verlag, New York, 1998.
12. J. Mathews and R. Walker, *Mathematical Methods of Physics*, Benjamin, New York, 1965.
13. P. M. Morse and H. Feshbach, *Methods of Theoretical Physics*, McGraw-Hill Book Company, New York, 1953.

14. J. F. Randolph, *Calculus*, Macmillan, New York, 1952.

15. J. R. Reitz, F. J. Milford, and R. W. Christy, *Foundations of Electromagnetic Theory*, 4th ed., Addison-Wesley, Reading, MA, 1993.

16. D. S. Saxon, *Elementary Quantum Mechanics*, Holden-Day, San Francisco, 1968.

17. G. N. Watson, *A Treatise on the Theory of Bessel Functions*, Cambridge University Press, Cambridge, 1922.

18. E. T. Whittaker and G. N. Watson, *A Course in Modern Analysis*, Cambridge University Press, Cambridge, 1927.

19. C. W. Wong, *Introduction to Mathematical Physics: Methods and Concepts*, Oxford University Press, New York, 1991.

Index